The Enigmatic Photon

Fundamental Theories of Physics

*An International Book Series on The Fundamental Theories of Physics:
Their Clarification, Development and Application*

Editor: ALWYN VAN DER MERWE
            University of Denver, U.S.A.

Editorial Advisory Board:

ASIM BARUT, *University of Colorado, U.S.A.*
BRIAN D. JOSEPHSON, *University of Cambridge, U.K.*
CLIVE KILMISTER, *University of London, U.K.*
GÜNTER LUDWIG, *Philipps-Universität, Marburg, Germany*
NATHAN ROSEN, *Israel Institute of Technology, Israel*
MENDEL SACHS, *State University of New York at Buffalo, U.S.A.*
ABDUS SALAM, *International Centre for Theoretical Physics, Trieste, Italy*
HANS-JÜRGEN TREDER, *Zentralinstitut für Astrophysik der Akademie der
    Wissenschaften, Germany*

The Enigmatic Photon

Volume 1: The Field $B^{(3)}$

by

Myron Evans
*Department of Physics,
University of North Carolina at Charlotte,
Charlotte, North Carolina, U.S.A.*

and

Jean-Pierre Vigier
*Department of Physics,
Université Pierre et Marie Curie,
Paris, France*

KLUWER ACADEMIC PUBLISHERS
DORDRECHT / BOSTON / LONDON

A C.I.P. Catalogue record for this book is available from the Library of Congress.

ISBN 0-7923-3049-8

Published by Kluwer Academic Publishers,
P.O. Box 17, 3300 AA Dordrecht, The Netherlands.

Kluwer Academic Publishers incorporates
the publishing programmes of
D. Reidel, Martinus Nijhoff, Dr W. Junk and MTP Press.

Sold and distributed in the U.S.A. and Canada
by Kluwer Academic Publishers,
101 Philip Drive, Norwell, MA 02061, U.S.A.

In all other countries, sold and distributed
by Kluwer Academic Publishers Group,
P.O. Box 322, 3300 AH Dordrecht, The Netherlands.

Printed on acid-free paper

All Rights Reserved
© 1994 Kluwer Academic Publishers
No part of the material protected by this copyright notice may be reproduced or
utilized in any form or by any means, electronic or mechanical,
including photocopying, recording or by any information storage and
retrieval system, without written permission from the copyright owner.

Printed in the Netherlands

Contents

PREFACE		ix
1.	**WAVE AND PARTICLE**	1
	1.1. The Enigma of Wave and Particle: Planck, Einstein, De Broglie	1
	1.2. Symmetry in Classical Electrodynamics, Wavenumber and Linear Momentum	6
	1.3. The Effect of $\boldsymbol{B}^{(3)}$ on the Fundamentals of the Old Quantum Theory	17
2.	**FUNDAMENTAL SYMMETRIES**	21
	2.1. The Seven Discrete Symmetries of Nature	21
	2.2. The $\hat{C}\hat{P}\hat{T}$ Theorem	24
	2.2.1. $\hat{P}$ Conservation	25
	2.2.2. $\hat{T}$ Conservation	26
	2.3. The Concept of Photon and Anti-Photon	26
	2.4. Parity of the Photon and Anti-Photon	28
	2.5. Motion Reversal Symmetry of Photon and Anti-Photon	29
	2.6. The Photon's Charge Conjugation Symmetry, $\hat{C}$	30
	2.7. Scheme for the Photon's Fundamental Symmetries, $\hat{C}$, $\hat{P}$ and $\hat{T}$	30
	2.8. Symmetry of the Pure Imaginary Field $-i\boldsymbol{E}^{(3)}/c$	33
	2.9. Experimental Demonstration of the Existence of the Anti-Photon	35

3. THE ORIGINS OF WAVE MECHANICS — 37

3.1. The Phase as Wave Function — 39
3.2. The Wave Mechanics of a Single Photon in Free Space — 40
3.3. Stationary States of One Photon in Free Space — 42
3.4. Heisenberg Uncertainty and the Single Photon — 46

4. INTER-RELATION OF FIELD EQUATIONS — 55

4.1. Relation Between the Dirac and D' Alembert Wave Equations — 55
4.2. Equations of the Quantum Field Theory of Light — 59
4.3. D'Alembert and Proca Equations — 65

5. TRANSVERSE AND LONGITUDINAL PHOTONS AND FIELDS — 71

5.1. Axial Unit Vectors, Rotation Generators, and Magnetic Fields — 72
5.2. Polar Unit Vectors, Boost Generators, and Electric Fields — 77
5.3. Lie Algebra of Electric and Magnetic Fields in the Lorentz Group, Isomorphism — 80
5.4. The Eigenvalues of the Massless and Massive Photons: Vector Spherical Harmonics and Irreducible Representations of Longitudinal Fields — 82

6. CREATION AND ANNIHILATION OF PHOTONS — 89

6.1. The Meaning of Photon Creation and Annihilation — 89
6.2. Quantum Classical Equivalence — 91
6.3. Longitudinal and Time-Like Photon Operators; Bilinear (Photon Number) Operators — 93
6.4. Light Squeezing and the Photomagneton $\hat{B}^{(3)}$ — 98

Contents vii

7. EXPERIMENTAL EVIDENCE FOR $\hat{B}^{(3)}$ — 103

- 7.1. The Inverse Faraday Effect, Magnetization by Light — 105
- 7.2. Optical NMR, First Order Effect of $B^{(3)}$ — 108
- 7.3. The Optical Faraday Effect (OFE) — 111
- 7.4. Survey of Data — 115

8. THE CONCEPT OF PHOTON MASS — 117

- 8.1. The Problem — 117
- 8.2. Brief Review of Experimental Evidence Compatible with $m_0 \neq 0$ and $B^{(3)} \neq 0$ — 120
- 8.3. The Proca Equation — 123
  - 8.3.1. Lorentz Invariance of the Maxwell Equations in Free Space — 125
  - 8.3.2. Covariance of the Proca Equation in Free Space — 126
- 8.4. Analogy between Photon Mass and Effective Current — 126
- 8.5. General Solutions of the Proca Equation — 129

9. AHARONOV-BOHM EFFECTS — 133

- 9.1. The Original Theory of Aharonov and Bohm — 134
- 9.2. The Effect of the Cyclic Algebra (25) — 136
- 9.3. The Optical Aharonov-Bohm Effect — 136
- 9.4. The Physical A_μ and Finite Photon Mass — 137
- 9.5. If the OAB Is Not Observed — 139
- 9.6. Non-Linearity of Photon Spin in Free Space — 142

10. MODIFICATIONS OF LAGRANGIAN FIELD THEORY — 147

- 10.1. Novel Gauge Fixing Term — 150
- 10.2. Quantization of the Electromagnetic Field — 152
- 10.3 A Potential Model for $B^{(3)}$ — 158

11. PSEUDO FOUR-VECTOR REPRESENTATIONS OF ELECTRIC AND MAGNETIC FIELDS — 161

11.1. Relation between the Minkowski and Lorentz Forces — 163
11.2. Dual Pseudo Four-Vector of $F_{\rho\sigma}$ in Free Space — 165
11.3. Link between V_μ, B_μ and E_μ — 166
11.4. Some Properties of E_μ and B_μ in Free Space — 167
11.5. Consequences for the Fundamental Theory of Free Space Electromagnetism — 169
11.6. Consequences for the Theory of Finite Photon Mass — 169

12. DERIVATION OF $B^{(3)}$ FROM THE RELATIVISTIC HAMILTON-JACOBI EQUATION OF e IN A_μ — 171

12.1. Action and the Hamilton-Jacobi Equation of Motion — 171
12.2. Solution of the Relativistic Hamilton-Jacobi Equation (369) — 174
12.3. The Orbital Angular Momentum of the Electron in the Field — 178
12.4. Limiting Forms of Equation (405) — 179
12.5. Discussion — 180

APPENDICES

A. Invariance and Duality in the Circular Basis — 185

B. Angular Momentum in Special Relativity — 189

C. Standard Expressions for the Electromagnetic Field in Free Space, with Longitudinal Components — 193

D. The Lorentz Force Due to $F^{(3)}_{\mu\nu}$; and $T^{(3)}_{\mu\nu}$ — 199

REFERENCES — 205

INDEX — 213

Preface

While investigating the theory of optically induced line shifts in N.M.R., one of us (MWE) came across the result that the antisymmetric part of the intensity tensor of light is directly proportional in free space to an entirely novel, phase free, magnetic field of light, which was identified as $B^{(3)}$, and which is defined in Eq. (4) of the first chapter of this book. The other chapters and development of the consequences of $B^{(3)}$ emerged after extensive consultation with Jean-Pierre Vigier and Keith Earle. The presence of $B^{(3)}$ in free space shows that the usual, propagating, transverse waves of electromagnetic radiation are linked geometrically to the spin field $B^{(3)}$, which indeed emerges directly (Chap. 12) from the fundamental, classical equation of motion of a single electron in a circularly polarized light beam. The field $B^{(3)}$ produces magnetization in an electron plasma which is proportional to the square root of the power density dependence of the circularly polarized electromagnetic radiation, conclusive evidence for the presence of the phase free $B^{(3)}$ in the vacuum. There are many experimental consequences of this finding, some of which are of practical utility, such as optical NMR, reviewed in Chap. 7, but the most important theoretical consequence is that there exist longitudinal components in free space of electromagnetic radiation, a conclusion which is strikingly reminiscent of that obtained from the theory of finite photon mass. The two ideas are interwoven throughout the volume. The characteristic square root light intensity dependence of $B^{(3)}$ dominates (Chap. 12), and is theoretically observable at low cyclotron frequencies when intense, circularly polarized electromagnetic radiation interacts with a single electron, or in practical terms an electron plasma or beam. The magnetization induced in such an electron ensemble by circularly polarized radiation is therefore expected to be proportional to the square root of the power density (i.e. the intensity in watts per square meter) of the radiation. This result emerges *directly* from the fundamental, classical, equation of motion of one electron in the beam, the relativistic Hamilton-Jacobi equation. To establish the physical presence of $B^{(3)}$ in the vacuum therefore requires the observation of this magnetization as a function of the beam's power density, a critically important experiment. Other possible experiments to detect $B^{(3)}$, such as the optical equivalent of the Aharonov-Bohm effect, are suggested throughout the volume.

We acknowledge many stimulating conversations with Dr. Keith Earle, and the authors are greatly indebted to Prof. Alwyn van der Merwe for the opportunity of producing this book in his prestigious and widely acclaimed series of monographs in fundamental physics. We also acknowledge the advice and suggestions of several colleagues who are interested in the magnetizing properties of circularly polarized light and electromagnetic radiation in general. A special word of thanks is due to the late Prof. Dr. Stanisław Kielich, who from the outset strongly encouraged the development of the theory of $B^{(3)}$.

Finally, we owe a great debt of gratitude to Dr. Laura J. Evans, whose highly professional production of the camera ready copy has been vital to the whole project.

Charlotte, NC, USA   MYRON W. EVANS
Paris, France   JEAN-PIERRE VIGIER
April, 1994

CHAPTER 1. WAVE AND PARTICLE

1.1 THE ENIGMA OF WAVE AND PARTICLE: PLANCK, EINSTEIN, DE BROGLIE

Light appears to be made of waves and simultaneously of particles, undulatory and particulate at the same time. This is an ancient enigma of natural philosophy, mysterious and implacable. The term "photon" is to be found in the dictionary [1] these days, defined as the quantum of electromagnetic energy, $h\nu$. Here h is Planck's constant and ν the frequency of an electromagnetic *wave*. We tell ourselves that there is a particle, the photon, which is also a wave, and which appears to have no mass. In the Galilean and Newtonian world this is not possible, reducing the mass of a Newtonian corpuscle to zero results in nothing at all. Energy in the Newtonian universe is simply an indivisible scalar, appearing to have nothing to do with frequency. The basis of wave mechanics, the de Broglie Guiding theorem [2, 3],

$$h\nu = m_0 c^2, \qquad (1)$$

for photons *and* for particles of matter, such as electrons, represents a world which is radically (i.e., at the root of things) different from the concepts enunciated by Galileo and Newton three or four centuries ago.

Since c is the speed of light and m_0 represents, after all, a mass in Eq. (1) our contemporary appreciation of light is a dark enigma of nature. We must reconcile ourselves to the fact that there is a particle without mass, whose energy is given by a frequency, and which can, after all, be defined as a mass multiplied by the square of the speed of light in vacuo, or free space. This particle is the photon and its concomitant wave produces the frequency ν. If we know a little about special relativity, we must reconcile ourselves to the fact that the speed c is a constant and the same in all Lorentz frames of reference, so that a particle travelling at c cannot exist in a frame in which it is at rest. An enigma indeed, making nonsense of common sense. Not only does this particle have energy, $h\nu$, we are told, but a linear momentum, $h\nu/c$, which is made up of two universal constants, h

and c, and a *wave frequency*, ν, so that linear momentum is $h\kappa/2\pi$, where κ, with the units of inverse distance (m^{-1}), is a wave-number. In our enigmatic world, the linear momentum of a massless particle is a wave-number. The angular momentum of this massless particle is fixed, and intrinsic, having eigenvalues $h/2\pi$ ($=\hbar$) and $-h/2\pi$ from considerations of special relativity. These eigenvalues are independent of frequency, and remain the same for photons of energy, $h\nu$, from radio frequencies to those of gamma rays. Hundreds of textbooks on the classical theory of electromagnetism assert with various degrees of certitude that all Maxwellian waves in vacuo are transverse, but the eigenvalues $\hbar$ and $-\hbar$ (being those of a boson with the value 0 missing) are *longitudinal* projections in the axis of propagation of the enigmatic photon or wave. Even in classical electrodynamics [4], the only non-vanishing eigenvalue of the angular momentum of light (transverse waves) is longitudinal in nature. The spin angular momentum of the electromagnetic plane wave is longitudinal in classical theory, and also independent of the frequency of the transverse wave.

The photon is therefore asserted to be a particle with no mass, but with a spin angular momentum, a linear momentum, and to be a light quantum of energy $h\nu$, defined in terms of a wave frequency ν. If $\boldsymbol{B}^{(1)}$ is the magnetic flux density of an electromagnetic plane wave, the light quantum is expressible as [5–8]

$$h\nu = \hbar\omega = \frac{1}{\mu_0}\int \boldsymbol{B}^{(1)} \cdot \boldsymbol{B}^{(1)*} dV, \qquad (2)$$

where μ_0 is magnetic permeability in vacuo and V a definite volume of space, so that $h\nu$ is also magnetic (or electric) in nature. In terms of the electric field strength of the plane wave and the electric permittivity, ϵ_0, of free space,

$$\hbar\omega = \epsilon_0 \int \boldsymbol{E}^{(1)} \cdot \boldsymbol{E}^{(1)*} dV. \qquad (3)$$

The light quantum is therefore both magnetic and electric in nature, and can be defined only with reference to a definite volume V in space. It is therefore not a point particle, and a pulse of electromagnetic radiation in free space behaves as a massless particle occupying a definite volume with fixed spin angular momentum, whose eigenvalues are projections in the axis of propagation. Yet the pulse of radiation is made up of transverse waves from the Maxwell equations [4]. These transverse waves in conventional understanding can have only two polarizations (right and left circular) and the correct premultiplier in Planck's radiation law of 1900, the first paper on which we shall shortly describe, is obtained by using the factor of two for the polarizations of standing waves in a cavity.

Yet, there is something profoundly unsatisfactory in this view of both classical and quantum electrodynamics because of the existence [9] of the cyclically symmetric relation between fields:

$$\boldsymbol{B}^{(1)} \times \boldsymbol{B}^{(2)} = iB^{(0)} \boldsymbol{B}^{(3)*}, \tag{4a}$$

in which (1), (2), and (3) can be permuted to give

$$\boldsymbol{B}^{(2)} \times \boldsymbol{B}^{(3)} = iB^{(0)} \boldsymbol{B}^{(1)*}, \tag{4b}$$

$$\boldsymbol{B}^{(3)} \times \boldsymbol{B}^{(1)} = iB^{(0)} \boldsymbol{B}^{(2)*}. \tag{4c}$$

Here, the fields $\boldsymbol{B}^{(1)}$, $\boldsymbol{B}^{(2)}$ and $\boldsymbol{B}^{(3)}$ are simply components of the magnetic flux density of *free space* electromagnetism in a circular, rather than in a Cartesian, basis. In the quantum field theory the longitudinal component $\boldsymbol{B}^{(3)}$ becomes the fundamental photomagneton of light, an operator defined by [10-15]

$$\hat{B}^{(3)} = B^{(0)} \frac{\hat{J}}{\hbar}, \tag{5}$$

where $\hat{J}$ is the angular momentum operator of one photon. The existence of the longitudinal $\hat{B}^{(3)}$ in free space is indicated *experimentally* by optically induced NMR shifts [16-18] and by several well known phenomena of magnetization by light, for example the inverse Faraday effect [19-26]. In this phenomenon, the phase free magnetization $\boldsymbol{M}^{(3)}$ can be deduced at second order by [18]

$$\boldsymbol{M}^{(3)} = AB^{(0)} \boldsymbol{B}^{(3)}, \tag{6}$$

where A is an ensemble averaged hyperpolarizability, a property of the material being magnetized by a circularly polarized laser beam.

The enigmatic photon therefore generates *three* degrees

of dimensionality in magnetic flux density in free space, labelled (1), (2) and (3) in the circular basis. Only two degrees of polarization are used customarily however in the derivation of the Planck radiation law, and classical limits thereof such as the Rayleigh-Jeans law [27]. In Bose's derivation of the Planck law, the concept of a particle was introduced [27] with only two degrees of polarization in three dimensional space. It is easily demonstrated however that the existence in free space of $\boldsymbol{B}^{(3)}$ is compatible with fundamental conservation laws in electrodynamics. For example, the following are Lorentz invariants in free space:

$$L_1 := F^{\mu\nu}F^*_{\mu\nu} = 2(c^2\boldsymbol{B}^{(i)} \cdot \boldsymbol{B}^{(i)*} - \boldsymbol{E}^{(i)} \cdot \boldsymbol{E}^{(i)*}) = 0, \tag{7}$$

$$L_2 := F^{\mu\nu}G^*_{\mu\nu} = 0, \quad (i) := (1), (2), (3).$$

For (i) = (1) and (2) Eqs. (7) reduce to the usual expressions [28] for the Lorentz invariants of the electrodynamic four-tensor $F_{\mu\nu}$ and its dual $G_{\mu\nu}$, expressed in a circular rather than a Cartesian basis. For (i) = (3), however, Eqs. (4) show that there is a real $\boldsymbol{B}^{(3)}$ which is dual [28] to the imaginary $-i\boldsymbol{E}^{(3)}/c$ (S.I. units). The latter, unlike the real $\boldsymbol{B}^{(3)}$, has no physical significance as a field, but the product $\boldsymbol{E}^{(3)} \cdot \boldsymbol{E}^{(3)*}$ is real, and therefore physical, and ensures that the Lorentz invariants L_1 and L_2 in Eq. (7) remain zero in free space for (i) = (3) as well as for (i) = (1) and (2).

It follows that the energy of one photon, the light quantum $\hbar\omega$ is as well described by (i) = (3) as by (i) = (1) or (i) = (2). In free space

$$\hbar\omega = \frac{1}{\mu_0}\int \boldsymbol{B}^{(i)} \cdot \boldsymbol{B}^{(i)*} dV = \epsilon_0 \int \boldsymbol{E}^{(i)} \cdot \boldsymbol{E}^{(i)*} dV, \tag{8}$$

$$(i) = (1), (2), (3).$$

The Poynting vector generated by the cross product of $\boldsymbol{B}^{(3)}$ and $-i\boldsymbol{E}^{(3)}/c$ is however always zero, because these quantities are always parallel vectors, so that the measured intensity of light (the time average of the Poynting vector) is not augmented by the presence of the physical $\boldsymbol{B}^{(3)}$ in free space.

The conventional derivation of the classical Rayleigh-Jeans law, using only (1) and (2), is correct only because the time averaged Poynting vector is unaffected by $\boldsymbol{B}^{(3)}$ and its dual, $-i\boldsymbol{E}^{(3)}/c$, in free space. More generally, the third

The Enigma of Wave and Particle

dimension (3) is needed for the enigmatic photon as much as for any other particle in three space dimensions. The phenomenon of magnetization by light described by Eq. (6) is a clear *experimental* corroboration of this deduction.

Critical scientific minds since the time of Cavendish and earlier have repeatedly come to the conclusion that light may have mass [29–35]. That this deduction affects electrodynamics at the most profound level (both in the classical and quantum theories) is exemplified by the fact that a pulse of electromagnetism in a vacuum would no longer be equivalent, in a finite volume V, to a particle, the photon, without mass. This leads to the replacement of the classical d'Alembert equation with a Proca equation [36],

$$\Box A_\mu = 0, \qquad (m_0 = 0), \tag{9a}$$

$$\downarrow$$

$$\Box A_\mu = -\left(\frac{m_0 c}{\hbar}\right)^2 A_\mu, \qquad (m_0 \neq 0), \tag{9b}$$

where A_μ is the electromagnetic potential four-vector as usual, and where m_0 is the mass of the photon in a rest frame, i.e., the photon *rest-mass*. In the d'Alembert equation there is no photon mass, and therefore no rest frame, so the structure of the equation remains invariant to Lorentz transformations from one frame of reference to another. This is no longer the case in the Proca equation, which is relativistically covariant, like all valid laws of physics, but no longer Lorentz invariant. In other words the photon with mass is a relativistic particle, whose velocity and mass varies from one Lorentz frame to another, even in free space.

Experimentally at present it appears possible only to put an upper limit on the photon mass [29], a limit which now replaces zero in standard tables, and which is thought to be in the range 10^{-68} to 10^{-45} kg. For all practical purposes in the laboratory, such a tiny figure means that the theory of light with mass is that of light without mass, yet there is a profound physical difference between the d'Alembert and Proca equations. On an astronomical scale, however, [29] finite m_0 implies measurable effects, for example in light reaching an earthbound observer from galaxies at the edge of the universe. The velocity of photons reaching earth from such sources should have slowed considerably below the universal constant c of special relativity. "Tired light" phenomena such as these have been reviewed in the literature of several scientific generations [29–35]. Clearly, the enigmatic photon does not fully reveal itself to the observer, but the constant c should not be regarded as an unchanging speed of light. It is a constant postulated in Einstein's second principle of special relativity. Photons with mass, however tiny, will not travel at c in free space.

In the de Broglie Guiding theorem, Eq. (1), the basis of wave mechanics in light and matter, the quantum of energy $\hbar\omega$ is equated with rest energy, being rest mass, m_0, multiplied by c^2, as deduced from special relativity. We are faced with the most profound enigma of all, that of representing mass as a frequency or vice versa. Contemporary gauge theory asserts [36] almost as an axiom that $m_0 =? 0$, leading to neat results such as those of Eq. (7). The existence of generations of scientific thought leading to $m_0 \neq 0$ for the photon, (or classical light) is considered only by a small minority of contemporary theoreticians [29–35]. However, these have succeeded in showing that $m_0 \neq 0$ for the photon can be at the least approximately compatible with the powerful results of unified and grand unified field theory [36], and with fundamental theorems of conservation of charge and current such as Noether's theorem [37]. In our present understanding, it is perhaps fair to say that both advantages and disadvantages appear to accrue from finite photon mass, the photon remaining an enigma.

1.2 SYMMETRY IN CLASSICAL ELECTRODYNAMICS, WAVENUMBER AND LINEAR MOMENTUM

Starting with the complex transverse vector potential [38]

$$\mathbf{A}^{(1)} = \mathbf{A}^{(2)*} = \frac{B^{(0)}}{\sqrt{2}\kappa}(i\mathbf{i} + \mathbf{j})e^{i\phi}, \qquad (10)$$

it follows that

$$\nabla \times \mathbf{A}^{(1)} = \kappa \mathbf{A}^{(1)} = \mathbf{B}^{(1)}, \qquad (11)$$

and that

$$\mathbf{B}^{(3)} = \mathbf{B}^{(3)*} = -\frac{i\kappa^2}{B^{(0)}}\mathbf{A}^{(1)} \times \mathbf{A}^{(2)}, \qquad (12)$$

showing that there is in principle a cyclic relation among vector potentials of the type (4) developed for magnetic fields in field space. However, the cyclic symmetry is not yet complete in the form of Eq. (12), and we establish the final relations later. However, Eq. (11) reveals a *symmetry duality* in the wavenumber κ, a duality which indicates that classical electrodynamics has its conceptual limits. Although κ as used in Eq. (11) must be a scalar, it mediates a direct proportionality between $\mathbf{B}^{(1)}$ (an axial vector positive to parity inversion $\hat{P}$ and negative to motion reversal $\hat{T}$) and $\mathbf{A}^{(1)}$ (which is $\hat{T}$ negative, $\hat{P}$ negative, a polar vector). Eq. (11) therefore suggests that the scalar κ is $\hat{P}$ negative, a property which is reminiscent of *vector linear momentum*. The scalar magnitude of the wave vector κ is therefore linked with particle momentum. *There is wave-particle dualism inherent in the classical Maxwell equations in free space.* The fact that the scalar κ is $\hat{P}$ negative indicates that the classical field is not a complete understanding. From Eq. (11)

$$\nabla \times (\nabla \times \mathbf{A}^{(1)}) = \kappa \nabla \times \mathbf{A}^{(1)} = \kappa^2 \mathbf{A}^{(1)}, \qquad (13)$$

i.e.,

$$\nabla^2 \mathbf{A}^{(1)} = -\kappa^2 \mathbf{A}^{(1)}, \qquad (14)$$

which is an eigenfunction equation reminiscent of the static limit of the Proca equation (9). Note, however, that: a) there is a change of sign between Eqs. (9) and (14); b) Eq. (14) conserves $\hat{P}$ and $\hat{T}$ symmetry, unlike Eq. (11), despite the fact that Eq. (14) is a direct consequence of Eq. (11), and so if the former is accepted as physical then so must the latter. The quantity κ has symmetry that cannot be understood in the framework of Maxwell's classical field equations.

The introduction of quantum concepts is one way of resolving the fundamental symmetry dualism inherent in Maxwell's equations, and this was first realized by Max Planck in November 1900 (Sec. 1.3).

The fact that Eq. (14) is an eigenfunction equation shows that ∇^2 plays the role of a quantum mechanical operator, whose eigenvalue is $-\kappa^2$. The vector potential $\mathbf{A}^{(1)}$ then plays the role of a wavefunction. From the fundamental axioms [39] of quantum mechanics

$$\hat{\nabla} = \frac{i}{\hbar} \hat{p}, \tag{15}$$

i.e., the del vector becomes a del operator, directly proportional by axiom to a linear momentum operator, $\hat{p}$. From Eq. (39) in Eq. (14)

$$\hat{p}^2 \mathbf{A}^{(1)} = \hbar^2 \kappa^2 \mathbf{A}^{(1)}. \tag{16}$$

Thus,

$$p^2 := \langle \hat{p}^2 \rangle = \hbar^2 \kappa^2, \quad p = \hbar \kappa = \hbar \frac{\omega}{c} = h \frac{v}{c}, \tag{17}$$

which defines the linear momentum of the photon in free space. Note that the Dirac constant is given by

$$\hbar = \frac{p}{\kappa}. \tag{18}$$

Therefore κ is identified as the expectation value of the operator $\hat{\kappa} = \hat{p}/\hbar$; i.e., the expectation value of a linear momentum operator of quantum field theory. However, κ is also a component of the classical wave vector, and so is an example of wave particle dualism.

The de Broglie Guiding theorem, Eq. (1), is derived in radiation by noting from Eq. (10) that

$$\frac{1}{c^2} \frac{\partial^2}{\partial t^2} \mathbf{A}^{(1)} = -\frac{\omega^2}{c^2} \mathbf{A}^{(1)} = -\kappa^2 \mathbf{A}^{(1)}. \tag{19}$$

The axioms of quantum mechanics imply that

$$\frac{\partial}{\partial t} = -\frac{i}{\hbar} En, \tag{20}$$

where En is an energy. From Eq. (20) in Eq. (19)

$$En = \kappa c \hbar = pc = m_0 c^2 = \omega \hbar = v h, \tag{21}$$

where m_0 is a mass defined formally by $p = m_0 c$. This is Eq. (1), whose structure is seen to be inherent in Maxwell's classical field equations. If there is considered to be finite photon rest mass, m_0, then the quantum of light energy hv is identified with the relativistic rest energy $m_0 c^2$ of the massive photon. If the photon mass is identified with zero axiomatically [36] as is often the case in contemporary field theory [37], then $h\nu$ is pure energy, i.e., is definable in terms only of energy and not of mass, despite the fact that it is formally identifiable with $m_0 c^2$.

From Eqs. (14) and (19)

$$\Box \mathbf{A}^{(1)} = \left(-\nabla^2 + \frac{1}{c^2}\frac{\partial^2}{\partial t^2}\right)\mathbf{A}^{(1)} = 0, \qquad (22)$$

which is the space part of the d'Alembert equation:

$$\Box A_\mu = 0, \qquad A_\mu := \epsilon_0\left(\mathbf{A}, \frac{i\phi}{c}\right). \qquad (23)$$

Equation (10) is a transverse solution of Eq. (23) and so Eq. (12) shows that $\mathbf{B}^{(3)}$ is formed from the cross product of complex conjugate solutions of the d'Alembert equation [40].

The question now arises as to what is the formal equivalent (in terms of vector potentials) of the novel cyclic relations (4). In other words, is there a cyclically symmetric relation between components of the vector potential in free space for a plane wave? If there is one, the longitudinal, phase free, component must be formed from the cross product of two polar vectors, $\mathbf{A}^{(1)}$ and its complex conjugate $\mathbf{A}^{(2)}$, and so must be an *axial* vector. For reasons developed later (Sec. 2.9), this axial vector can be written as the pure imaginary $i\mathbf{A}^{(3)}$. The formal cyclic relation akin to Eqs. (4) is therefore:

$$\mathbf{A}^{(1)} \times \mathbf{A}^{(2)} = -A^{(0)}(i\mathbf{A}^{(3)})^*, \qquad \mathbf{A}^{(2)} \times (i\mathbf{A}^{(3)}) = -A^{(0)}\mathbf{A}^{(1)*},$$
$$(i\mathbf{A}^{(3)}) \times \mathbf{A}^{(1)} = -A^{(0)}\mathbf{A}^{(2)*}, \qquad (24)$$

in which the permuted vectors are $\mathbf{A}^{(1)}$, $\mathbf{A}^{(2)}$ and $i\mathbf{A}^{(3)}$. These are relations among the space components of a *complex* four-vector A_μ [41]. There is therefore a set of three cyclically symmetric structures between plane wave components in free space:

$$\mathbf{B}^{(1)} \times \mathbf{B}^{(2)} = iB^{(0)}\mathbf{B}^{(3)*}, \qquad \text{and cyclic permutations,} \qquad (25a)$$

$$\mathbf{A}^{(1)} \times \mathbf{A}^{(2)} = -A^{(0)}(i\mathbf{A}^{(3)})^*, \qquad \text{and cyclic permutations,} \qquad (25b)$$

$$\boldsymbol{E}^{(1)} \times \boldsymbol{E}^{(2)} = -E^{(0)}(i\boldsymbol{E}^{(3)})^*, \text{ and cyclic permutations,} \qquad (25c)$$

in which the longitudinal *magnetic* component is physical and pure real, while $i\boldsymbol{A}^{(3)}$ and the electric field $i\boldsymbol{E}^{(3)}$ are pure imaginary and unphysical at first order. All three longitudinal ((3)) components are *phase free* because they are formed from the cross product of two complex conjugates. Under the dual transformation of special relativity, we have, in Minkowski's notation [42] (in which $x_\mu := (X, Y, Z, ict)$)

$$\boldsymbol{B} \to -\frac{i}{c}\boldsymbol{E}, \qquad \boldsymbol{E} \to ic\boldsymbol{B}, \qquad (26)$$

so that the scalar amplitudes transform in free space according to

$$B^{(0)} \to -iB^{(0)}, \qquad E^{(0)} \to iE^{(0)}, \qquad A^{(0)} \to -iA^{(0)}, \qquad (27)$$

leaving Maxwell's equations invariant. Applying Eqs. (27) to Eqs. (25) the following results are obtained.

(1) Each of equations (25) is unchanged under the dual transformation, i.e., each is *self-dual*. Each is relativistically invariant in free space and formally covariant [41, 42].

(2) Using the free space dual transformations:

$$B^{(0)} \to -i\frac{E^{(0)}}{c}, \qquad E^{(0)} \to icB^{(0)},$$
$$A^{(0)} \to -i\frac{B^{(0)}}{\kappa} = -i\frac{E^{(0)}}{\omega}, \qquad (28)$$

it is seen that each equation of the set (25) is self-dual and also dual with the other two. For example, the transformation $B^{(0)} \to -iE^{(0)}/c$ converts Eq. (25a) into Eq. (25c). The dual transformation $A^{(0)} \to -iB^{(0)}/\kappa$ converts Eq. (25b) into Eq. (25a) and so on.

(3) These properties are lost in the conventional picture of free space electrodynamics, in which there is no phase free spin field $\boldsymbol{B}^{(3)}$, only the phase dependent wave fields $\boldsymbol{B}^{(1)}$ and $\boldsymbol{B}^{(2)}$. In the conventional view [43], $\boldsymbol{B}^{(3)} =?\ \boldsymbol{0}$, so that the Lorentz invariance of Eq. (4) is lost. For example, if $\boldsymbol{B}^{(1)} \times \boldsymbol{B}^{(2)} =?\ \boldsymbol{0}$, then

under the dual transform $B^{(1)} \times B^{(2)} \to -B^{(1)} \times B^{(2)}$ there is a change of sign, but the same transform results in $0 \to 0$, in which there is no change of sign. Therefore the assertion $B^{(3)} =? 0$ is relativistically inconsistent in free space, and from Eq. (4) is also of course algebraically incorrect. It is also inconsistent to assert that the imaginary $iE^{(3)}$ and $iA^{(3)}$ are zero, they are non-zero and pure imaginary.

(4) Finally, the Lorentz invariants of Eq. (7) are self-dual, and by definition are not affected by a Lorentz transformation.

We find that one of the basic tenets of electrodynamics in free space is incomplete, there exists a real and physical spin field $B^{(3)}$ which can magnetize matter as in the inverse Faraday effect [19–26]. In what way does this very recent realization [10–15] affect the fundamental papers of the quantum theory, due to Planck, Einstein, Bose and their contemporaries? This question is partially answered through a consideration of the time averaged Poynting vector, or intensity of light in watts per square meter, the total electromagnetic energy per unit time per unit area, or energy flux density at a temperature T. This quantity is expressed as an integral over frequency,

$$I(T) = \int_0^\infty I_\nu(T) \, d\nu. \qquad (29)$$

The total electromagnetic energy density (U) is related to the intensity I by

$$U = \frac{4}{c} I. \qquad (30)$$

Both U and I are unchanged by $B^{(3)}$, because it is generated by the *spin* of the photon and cannot contribute to the time averaged Poynting vector [44]. It follows that $B^{(3)}$ does not contribute to the classical Rayleigh-Jeans law, a limit of the Planck radiation law, the essence of which is Planck's hypothesis: that the total electromagnetic energy at a particular frequency is the sum of identical energy elements, or light quanta $\hbar\omega$. Since $B^{(1)} \cdot B^{(1)*} = B^{(2)} \cdot B^{(2)*} = B^{(3)} \cdot B^{(3)*}$, the light quantum $\hbar\omega$ can be expressed equivalently as an integral over any of these dot products. The physical explanation is that each dot product represents an *average*.

The wave fields $\mathbf{B}^{(1)}$ and $\mathbf{B}^{(2)}$ depend on phase, $\phi = \omega t - \mathbf{\kappa} \cdot \mathbf{r}$, where t denotes an instant in time and $\mathbf{r}$ a position in space. The spin field $\mathbf{B}^{(3)}$ does not. However, the factor $B^{(0)}$, the scalar amplitude defined by

$$B^{(0)} = \left(\frac{\mu_0 \hbar \omega}{V}\right)^{\frac{1}{2}}, \qquad (31)$$

is common to all three fields in free space, so the *amplitudes* of all three fields follow the Planck distribution. Since $\mathbf{B}^{(3)}$ *is not a wave*, i.e., does not vary with phase, ϕ, and is not associated with any given frequency ω, or wave vector $\mathbf{\kappa}$, it cannot be absorbed by an atom or molecule in the same way as $\mathbf{B}^{(1)}$ and $\mathbf{B}^{(2)}$. The field $\mathbf{B}^{(3)}$, furthermore, can magnetize material far from optical resonance, as in the inverse Faraday effect [19–26]. This process of magnetization can occur, in other words, without resonant absorption, e.g. using laser frequencies far removed from any natural optical absorption mode of the atom or molecule.

Panofsky and Phillips [45] for example point out that integrals of the type

$$\hbar \omega = \frac{1}{\mu_0} \int \mathbf{B}^{(i)} \cdot \mathbf{B}^{(i)*} dV, \qquad (i) = (1), (2), (3), \qquad (32)$$

must have the same transformation properties of *point mass* in special relativity. This finding, at first strange and counter-intuitive, is nevertheless consistent with the de Broglie Guiding theorem, Eq. (1), which defines point mass, m_0 as $h\nu/c^2$. The amplitude $B^{(0)}$ is therefore tied to a mass of radiation in volume V. This picture of mass is fundamental to the general theory of relativity [45]. Considering, following Panofsky and Phillips [45], a volume V containing totally a quantity of electromagnetic radiation and no charges and currents, then the linear momentum and energy of a radiation pulse contained within that finite volume has the same transformation properties as a point particle, identified as *the photon*. Thus, the energy of one photon is $h\nu$, its linear momentum is $h\nu/c$, and its angular momentum, to which the spin field $\mathbf{B}^{(3)}$ is directly proportional, is $\pm\hbar$. The eigenvalues of the operator $\hat{B}^{(3)}$ are therefore $\pm B^{(0)}\hbar$.

For a plane electromagnetic wave, if W denotes its energy and $\mathbf{G}$ its linear momentum, the four-vector product $W^2 - c^2 G^2 = G^\mu G_\mu = 0$ is a Lorentz invariant, and the quantity

[45] (in covariant-contravariant notation),

$$G^\mu = \left(\int c\boldsymbol{G} dV, \int W dV\right), \tag{33}$$

is a four-vector (a linear momentum, energy four-vector of special relativity). There is no simple equivalent four-vector of angular momentum, angular energy because the angular momentum, being an axial vector, cannot form the space part of a four-vector from fundamental geometrical considerations [45]. Thus $\boldsymbol{B}^{(3)}$ (proportional to electromagnetic angular momentum density) and $\boldsymbol{B}^{(3)} \cdot \boldsymbol{B}^{(3)*}$ (proportional to electromagnetic energy density) cannot form a four-vector in the same way as the Poynting vector $\boldsymbol{N}$ and energy density $\boldsymbol{B}^{(i)} \cdot \boldsymbol{B}^{(i)*}$ (i) = (1), (2), (3) form four-vectors, and therefore do not affect fundamental relativistic relations such as (33) in free space. Since $G^\mu G_\mu$ is zero, it is light-like, and corresponds to zero mass [45], and electromagnetic radiation propagating with velocity c obeys *particle* transformation laws yielding finite momentum and energy.

Barut [46] gives a clear discussion of the conceptual consequences of a zero mass particle. This cannot exist in Newtonian mechanics. In special relativity, the definition of a four-momentum as

$$p_\mu = m_0 c v_\mu, \tag{34}$$

where v_μ is the four-velocity also breaks down if the mass is zero. For a zero mass particle the energy and linear momentum must become *primary concepts*. The Newtonian velocity and trajectory cannot be defined if mass is zero, and momentum cannot be connected to velocity. Localization of the massless particle is not defined or possible. Linear momentum can be defined only through the light-like four-vector (En/c, $\boldsymbol{p}$), with $En = pc$. For the photon, we have $En = \hbar\omega$ and $|\boldsymbol{p}| = \hbar\omega/c$. There is no *non-relativistic limit to this concept*. Similarly, the massless particle cannot have a non-zero Newtonian moment of inertia, and so the angular momentum of the particle (considered to have finite radius) cannot be related to its angular velocity. We see this through the fact that the angular momentum of the photon is the unvarying $\hbar$, while its variable angular velocity is the angular frequency ω. There is no linear relation of the type $\hbar = ? I_p \omega$, where I_p is a hypothetical photon moment of inertia. However, the product $\hbar\omega$ is an energy, in the same way as the classical mechanical product of angular momentum and angular velocity is a classical angular energy.

Furthermore, Barut points out that an electromagnetic wave is always characterized by the light-like four-vector $\kappa_\mu = (\omega/c, \kappa)$, with $\omega = \kappa c$. It follows that the newly discovered spin field $B^{(3)}$ [10–15] cannot be associated with linear momentum or angular frequency because it has no phase dependence. This is of course consistent with the fact that $B^{(3)}$ is proportional to photon angular *momentum* $\hbar$, which has no intrinsic frequency dependence. The latter is itself unacceptable in Newtonian mechanics, because an angular momentum is within that view proportional always to an angular frequency, whereas $\hbar$ is a universal constant, with the units of angular momentum, *independent* of angular frequency. The Planck hypothesis of 1900 is therefore relativistic in nature, although it was left to Lorentz, Poincaré and Einstein (1903-1905) to realize fully the nature of the light quantum hypothesis, as we shall shortly describe.

It is clear however that the newly discovered free space spin field $B^{(3)}$ also has no non-relativistic counterpart if the photon mass is taken to be zero. It is directly proportional to the photon angular momentum, the angular momentum of an (axiomatically) massless particle which cannot be Newtonian, and cannot be localized. Following Barut [46], when dealing with questions of energy and linear momentum interchange between a photon and other particles, such as electrons, the plane *waves* $B^{(1)}$ and $B^{(2)}$ of Maxwellian electrodynamics are associated with a massless particle. The *spin field* $B^{(3)}$ is associated with photon angular momentum, which in the inverse Faraday effect, for example, is transferred elastically to matter far from any optical resonance, i.e., far from any spectral frequency at which exchange of energy and linear momentum occurs through the phase. This is another way of saying that $B^{(3)}$ is phase free.

The following useful rules can therefore be constructed:

(1) When there is exchange of photon energy and linear momentum, through the phase, the Lorentz invariant product $\kappa^\mu x_\mu$, the *wave* fields $B^{(1)}$ or $B^{(2)}$ are involved.

(2) When there is exchange only of photon angular momentum, the *spin* field $B^{(3)}$ is involved.

The reduced Planck constant, $\hbar$, (also known as the Dirac constant), can therefore be thought of as a proportionality constant between energy and frequency, a constant which must be the same as that which mediates linear momentum and wavenumber,

$$\hbar = \frac{En}{\omega} = \frac{|\mathbf{P}|}{|\mathbf{\kappa}|} := \frac{P}{\kappa}, \qquad (35)$$

so that En and $\mathbf{p}$ defined for the plane wave satisfy [46] the energy and momentum relationship for a relativistic particle of zero mass. If the particle has mass, the *de Broglie wave particle dualism* implies that the particle is associated with a plane wave known as a matter wave, whose frequency ω and wave-vector are again given, following Barut [46] by equation (35), but in which κ_μ is no longer a light-like vector. Its relativistic energy is that of a particle with mass,

$$En^2 = c^2 p^2 + m_0^2 c^4, \qquad (36)$$

so that the relation between frequency (ω) and wavenumber becomes

$$\omega^2 = c^2 \kappa^2 + \frac{m_0^2 c^4}{\hbar^2}, \qquad (37)$$

as first proposed by de Broglie [47] and verified experimentally [46]. The relation (37) is consistent with the well known Proca equation (9b), in which the photon mass is of course non-zero. It can be seen that the photon mass, if non-zero, will change the optical relation between wave number and frequency, which will become different in different Lorentz frames. One of the interesting consequences [10–15] of finite photon mass is that the spin field $\mathbf{B}^{(3)}$ becomes

$$\mathbf{B}^{(3)} = B^{(0)} \exp\left(-\frac{m_0 c}{\hbar} Z\right) \mathbf{k}, \qquad (38)$$

and decays exponentially. Since m_0 is different in different Lorentz frames, the spin field (and also the wave fields) will also become different, and no longer Lorentz invariant as for the massless photon.

The constant $\hbar$ though, remains the same for zero or

non-zero photon mass, and is given by

$$\hbar = \frac{V}{\mu_0 \omega} B^{(0)2}. \tag{39}$$

From our above considerations, this is a relativistic quantum classical equivalence. For a photon without mass it has no non-relativistic limit, and it essentially describes the classical equivalent (right hand side) of the angular momentum *magnitude* of one photon in free space. This magnitude is $\hbar$, an unchanging property both of the massless and massive photon. Therefore if the right hand side is a classical one photon angular momentum magnitude, the classical angular momentum *vector* for one photon is

$$\boldsymbol{J}^{(3)} = \hbar \boldsymbol{k} = \left(\frac{V}{\mu_0 \omega}\right) B^{(0)2} \boldsymbol{k}, \tag{40}$$

where $\boldsymbol{k}$ must be an *axial* unit vector. We note that the vector defined by

$$\boldsymbol{B}^{(3)} := B^{(0)} \boldsymbol{k}, \tag{41}$$

has all the known properties of magnetic flux density (tesla) and is the spin field defined in equations (4). It follows immediately that

$$\boldsymbol{B}^{(3)} = \frac{\mu_0 \omega}{B^{(0)} V} \boldsymbol{J}^{(3)} = B^{(0)} \frac{\boldsymbol{J}^{(3)}}{\hbar}, \tag{42}$$

i.e., the classical $\boldsymbol{B}^{(3)}$ vector can be written as the product of the amplitude $B^{(0)}$ with the classical ratio $\boldsymbol{J}^{(3)}/\hbar$, which is the axial unit vector $\boldsymbol{k}$. This result emerges *directly* from the fundamental definition (39) with the basic definition $\boldsymbol{J}^{(3)} := \hbar \boldsymbol{k}$ which follows from the fact that $\hbar$ has the units of an angular momentum magnitude. From the definition of $\hbar$ in Eq. (35) it follows that the angular momentum of one photon in free space can be expressed as the classical vector

$$\boldsymbol{J}^{(3)} = \frac{En}{\omega} \boldsymbol{k} = \frac{p}{\kappa} \boldsymbol{k}, \tag{43}$$

which is the well known result from classical electrodynamics [4, 45],

$$J^{(3)} = \frac{\epsilon_0}{\kappa}\int E^{(1)} \times B^{(2)}\, dV = \frac{p}{\kappa}k. \qquad (44)$$

The spin field $B^{(3)}$ is therefore present in classical electrodynamics, as revealed in another way by Eqs. (4). Remarkably, these relations were first identified only in 1992 [9–15]. The quantum field theoretical equivalent of Eq. (42) is clearly [10]

$$\hat{B}^{(3)} = B^{(0)}\frac{\hat{J}^{(3)}}{\hbar}, \qquad (45)$$

where $\hat{B}^{(3)}$ and $\hat{J}^{(3)}$ are operators whose expectation values are simply $B^{(3)}$ and $J^{(3)}$.

If it is asserted [48] that $B^{(3)} =? 0$, the structure of electrodynamics is violated.

1.3 THE EFFECT OF $B^{(3)}$ ON THE FUNDAMENTALS OF THE OLD QUANTUM THEORY

The core logic of Eqs. (4) asserts that there exists a novel cyclically symmetric field algebra in free space, implying that the usual transverse solutions of Maxwell's equations are tied to the longitudinal, non-zero, real, and physical magnetic flux density $B^{(3)}$, which we name the spin field. This deduction changes fundamentally our current appreciation of electro-dynamics and therefore the principles on which the old quantum theory was derived, for example the Planck law [48] and the light quantum hypothesis proposed in 1905 by Einstein. The belated recognition of $B^{(3)}$ [9–15] implies that there is a magnetic field in free space which is associated with the longitudinal space axis, Z, which is labelled (3) in the circular basis. Conventionally, the radiation intensity distribution is calculated using only two, transverse, degrees of freedom, right and left circular, corresponding to (1) and (2) in the circular basis. The latter is therefore the natural basis for the consideration of electromagnetism in free space, and in terms of Cartesian unit vectors i, j, and k is defined by the circular unit vectors $e^{(1)}$, $e^{(2)}$, and $e^{(3)}$ such that

$$e^{(1)} = \frac{1}{\sqrt{2}}(i - ij), \quad e^{(2)} = \frac{1}{\sqrt{2}}(i + ij), \quad e^{(3)} = k. \qquad (46)$$

These circular unit vectors form the geometrical basis for the cyclical relations (4), i.e.,

$$\boldsymbol{e}^{(1)} \times \boldsymbol{e}^{(2)} = i\boldsymbol{e}^{(3)}, \qquad (47)$$

and cyclical permutations of (1), (2), and (3).

The existence of $\boldsymbol{B}^{(3)}$ is strikingly consistent with the fact that the Proca equation (9b) allows longitudinal solutions in free space. Its recognition is therefore consistent with finite photon mass. The fields $\boldsymbol{B}^{(3)}$ from the d'Alembert and Proca equations are for all practical purposes identical in the laboratory, because the photon mass is tiny. Finite photon mass means that electromagnetic waves behave in the same way as de Broglie matter waves, and that the photon becomes a properly defined boson, with eigenvalues $-\hbar$, 0, and $\hbar$, not $-\hbar$ and $\hbar$ as in the conventional theory of photon spin [15]. In classical, relativistic field theory, the Wigner little group for the massless photon is the unphysical Euclidean E(2), the group of rotations and translations in a plane [42]. This makes no sense in three dimensional Euclidean space, and standard texts in field theory simply note that this is an obscure facet of electromagnetic theory. The incorporation of photon mass, however tiny, improves matters, and allows a straightforward quantization [42]. It is also possible [49] to incorporate photon mass into unified field theory, and grand unified field theory, and there is no reason to assert on the grounds of gauge invariance that the photon is massless. The newly discovered existence of $\boldsymbol{B}^{(3)}$ means that the conventional "two dimensional" approach to electrodynamics must be abandoned. There is no purpose in accepting a theory which leads to an unphysical conclusion, such as the E(2) catastrophe of special relativity [42]. The spin field $\boldsymbol{B}^{(3)}$ is unmistakable evidence that the conventional theory is incompletely understood, both for finite and zero photon mass.

Does the existence of $\boldsymbol{B}^{(3)}$ mean that the laws of radiation must be modified? Should the value of the Planck constant be increased by a factor 3/2 to account for three dimensions rather than two?

The answer is in the negative, because $\boldsymbol{B}^{(3)}$ is phase free, and is not a wave field. It is a spin field, due to the photon spin, which is longitudinal and relativistically invariant. The derivation by Rayleigh [50] of the classical intensity law, a derivation corrected by Jeans [50], is based on the existence of two transverse standing wave components (left and right) in a cavity. The field $\boldsymbol{B}^{(3)}$ is not a wave and does not augment the radiation intensity. This conclusion is consistent with the fact that $\boldsymbol{B}^{(3)}$ does not augment

The Effect of $B^{(3)}$ on Quantum Theory

the Poynting vector, whose time average is radiation intensity in watts per unit area.

The law of radiation intensity due to Planck requires the assumption that there exist energy elements which are proportional to frequency, ν, through the Planck constant h,

$$\epsilon = h\nu = \hbar\omega. \tag{48}$$

As we have seen, the light quantum $h\nu$ can be expressed equally well (Eq. (8)) in terms of wave fields and spin field, but since the latter is proportional to photon angular momentum it cannot describe optical resonant absorption. Magnetization due to $B^{(3)}$ in the inverse Faraday effect can take place far from optical resonance, and is a process involving transfer of photon angular momentum. The absorption process described by

$$En_m - En_n = \hbar\omega, \tag{49}$$

needs wave fields with a phase dependence. Here En_n and En_m are energy levels of states n and m respectively of an atom. Absorption of light is a phase dependent process, i.e., depends on the four-vector κ_μ. The spin field $B^{(3)}$ has no phase dependence, like photon angular momentum, and so cannot be involved in a process of light absorption that depends on phase. The spin field $B^{(3)}$ can therefore be regarded as a wave field whose phase is zero, but whose amplitude $B^{(0)}$ is frequency dependent through Eq. (31). This amplitude, however, is the same for the wave fields $B^{(1)}$ and $B^{(2)}$ and for the spin field $B^{(3)}$.

The conventional picture of absorption by a photon:

$$En_m - En_n = h\nu, \qquad \frac{A_{mn}}{B_{mn}} = \rho(\nu, T)\left(e^{(En_m - En_n)/kT} - 1\right), \tag{50}$$

where En_m and En_n are atomic energy levels of states m and n respectively, and where A_{mn} and B_{mn} are Einstein coefficients, is therefore unchanged by the spin field $B^{(3)}$. In Eq. (50), $\rho(\nu)$ is the density of states as usual.

The derivation by Bose [50] of the Planck radiation law, without recourse to classical electrodynamics, was made by replacing the number of standing waves in a cavity of volume V with the counting of cells in a one particle position/momentum phase space dxdp. If there is a spin field or "field without phase frequency" $B^{(3)}$, then we see immediately that there can be no additional cells to count, and so the existence of $B^{(3)}$ does not change the validity of Bose's derivation of Planck's law, an important step towards the emergence of the notion of the photon as a particle. This is simply a way of recognizing that a wave must by definition have a phase dependence, and so $B^{(3)}$ is not a wave field. The essence of Bose's method is therefore to integrate the one particle phase space element dxdp over the volume V and over all linear momenta between p and $p+dp$, and to supply a factor of two to count polarizations and to arrive at the correct premultiplier in the Planck law. Why this factor is two was not known at the time of the derivation (1924) [50], essentially because the relativistically correct description of angular momentum was not known for a massless particle. The accepted treatment did not emerge until 1939, in a paper by Wigner [51] in which, however, the Wigner little group E(2) is a logical but physically obscure [42] consequence.

The concept of the photon as massless particle is therefore filled with enigma from the very outset, and by no means has the photon lost its power to surprise. The existence of a spin property in light was realized experimentally as long ago as 1811 by Arago in the form of left and right circular polarization, signalling the existence of (1) and (2) in our circular basis described by Eq. (46). The concept of longitudinal angular momentum (Eq. (44)) is present, again, in the classical Maxwellian description, and occurs as photon spin in the quantum theory. The only non-zero components of this spin are longitudinal, i.e., occur in the Z, or (3), axis. Finally, 181 years after Arago's realization, the field $B^{(3)}$ emerges through Eq. (4a). The third axis (3) in the circular basis is finally brought into consideration.

CHAPTER 2. FUNDAMENTAL SYMMETRIES

In writing the novel cyclical relations (4) it is assumed implicitly that there is no violation of the fundamental discrete symmetries [52] of nature: parity inversion ($\hat{P}$); motion reversal ($\hat{T}$) and charge conjugation ($\hat{C}$). A valid mathematical description of a natural phenomenon is objective, such an equation represents a law of physics, a law which is a necessary and sufficient description of the complete experiment. If an equation is unchanged by application of a discrete symmetry operator, e.g. $\hat{P}$, the law of physics is unchanged under parity inversion of the complete experiment and parity is conserved in the natural phenomenon described by the law. If not, parity is violated. The law of physics describing a complete experiment is embodied in an equation through which the experiment and the variables relevant to observation are defined. For example, in Newton's Law, force is the product of mass and linear acceleration, ($\boldsymbol{F} = m\boldsymbol{\dot{v}}$), and force is necessarily and completely defined by the two observables m and $\boldsymbol{\dot{v}}$. Any other description of force is subjective, and outside the boundaries of natural philosophy.

There have been suggestions [53] to replace this necessarily mathematical description of nature by the use of simple diagrams based purely on symmetry. While these may produce the correct mathematical result, they may as often lead to spurious conclusions, because without knowledge of the appropriate law of physics, it is not possible to identify objectively the relevant physical variables. The latter in the diagrammatic approach [53] to complete experiment symmetry have to be chosen subjectively, or arbitrarily, and if this is done without recourse to experimentation, or empiricism, the result is also arbitrary. Symmetry must therefore be used wherever possible with an established mathematical description.

2.1 THE SEVEN DISCRETE SYMMETRIES OF NATURE

The three fundamental symmetries $\hat{P}$, $\hat{T}$, and $\hat{C}$ can be combined to form the operators $\hat{P}\hat{T}$, $\hat{T}\hat{C}$, $\hat{P}\hat{C}$, and $\hat{C}\hat{P}\hat{T}$, giving a total of seven [52]. There are several specialist accounts in the literature [52] to which we refer the interested reader for the elements of symmetry theory. In electrodynamics (classical and quantum) we are concerned with the symmetries of electric and magnetic fields, charges and currents, in free space and material matter. Electrodynamics is based on the concepts of elementary charge and field. In contemporary terms, the field is a gauge field, denoted by the potential four-vector A_μ in the 4-D of special relativity [54]. Electromagnetism enters gauge theory, however, through the *product* eA_μ, having the dimensions of $\partial/\partial x_\mu$. The introduction of eA_μ is necessary to maintain the invariance of the appropriate Lagrangian [54] under type two gauge transformations, this invariance being regarded as fundamentally necessary, in the same way as conservation of charge, energy and momentum. The existence of the scalar constant e is a fundamental assumption of the theory as presently formulated, and e is identified as the charge on the electron

$$e = -1.60210 \times 10^{-19} \, C. \tag{51}$$

In electrodynamics, this unit of charge is conventionally defined as *negative definite*, i.e., in theories of electric and magnetic fields, e does not change sign unless the electron becomes a distinct (i.e., different) particle, the positron. The transmutation of electron to positron is one example of the action of the operator $\hat{C}$,

$$\hat{C}(e^-) = e^+, \tag{52}$$

which is therefore defined in elementary particle physics by the statement [52] that it generates the corresponding antiparticle from the original particle.

To apply this definition to classical electrodynamics requires care. From our introductory considerations, it follows that both sides of an equation of electrodynamics must have the same overall $\hat{C}$ symmetry, otherwise $\hat{C}$ is violated. However, the application of $\hat{C}$ changes the sign of the elementary e and so *transforms electrodynamics into a subject based on a positive elementary charge*. Similarly, $\hat{C}$ transforms the electrons of quantum electrodynamics into positrons, and simultaneously transforms matter into antimatter. Since $\hat{C}$, by definition [52] cannot affect spatiotemporal quantities such as $\partial/\partial x_\mu$, it follows that [52]

$$\hat{C}(A_\mu) = -A_\mu. \tag{53}$$

The gauge field described by A_μ becomes a gauge *anti*-field, interacting with anti-matter, made up of positrons, anti-protons, anti-neutrons and so forth. Appropriate initial, or boundary, conditions have however, assured that the earth, solar system, and presumably the universe as now understood, are naturally composed of matter, and not anti-matter, of fields and not anti-fields. Evolution has resulted in this condition, so that a description of naturally occurring fields and matter must be one based *on a negative definite e, i.e., on an e whose sign cannot be changed*. It follows that the scalar amplitude of A_μ, denoted $A^{(0)}$, *must also have a definite sign* attached to it in this description, the description known as electrodynamics.

Conventionally, the sign of A_μ is positive, so $A^{(0)}$ is positive definite, just as e is negative definite. Similarly, the scalar magnetic flux density amplitude $B^{(0)}$ and electric field strength amplitude $E^{(0)}$ of electromagnetism are positive definite. If we are to remain within the structure of electrodynamics, therefore, the signs of scalars such as $A^{(0)}$, $B^{(0)}$ and $E^{(0)}$ cannot change, signifying that $\hat{C}$ is an operator which takes us outside the limits of the subjects of classical and quantum electrodynamics *as applied to the natural, or evolved, universe.* The following examples illustrate this statement.

(1) Oscillating wave fields such as $\boldsymbol{B}^{(1)}$ and $\boldsymbol{E}^{(1)}$ in free space or matter (i.e., the complete vectors) must change sign at first order when oscillating, but this occurs through the oscillatory phase factor $e^{i\phi} = \cos\phi + i\sin\phi$ and not through the scalar pre-multipliers $B^{(0)}$ and $E^{(0)}$. The phase is spatio-temporal in nature, and by definition is unaffected by $\hat{C}$.

(2) Applying $\hat{C}$ does not produce, for example, a cation from an anion, since both are made of electrons, protons and neutrons of *matter*. ($\hat{C}$ must produce anti-matter, i.e., positrons, anti-protons and anti-neutrons.) Similarly, applying $\hat{C}$ to a macroscopically charged object, such as a van der Graaf generator, results in an object made of anti-matter, an object which is not observed naturally, since anti-matter apparently does not occur in the natural universe.

(3) Free space electromagnetism is considered to

emerge from the product eA_μ, and therefore from a source made up of radiating electrons located in a universe of matter and fields. Currents in this source are necessarily moving electrons. Free space electromagnetism therefore arises conventionally from matter at a source in the natural universe and never from anti-matter. Applying $\hat{C}$ to A_μ means that its source must become anti-matter.

It is clear that the novel Eqs. (4) of chapter (1), the equations that tie together the spin field $\boldsymbol{B}^{(3)}$ and wave fields $\boldsymbol{B}^{(1)}$ and $\boldsymbol{B}^{(2)}$, conserve $\hat{C}$, because both sides are $\hat{C}$ positive, both sides being to order $B^{(0)2}$. In coming to this conclusion, we note that

$$\hat{C}(A^{(0)}) = -A^{(0)}, \qquad \hat{C}(B^{(0)}) = -B^{(0)}, \qquad \hat{C}(E^{(0)}) = -E^{(0)}, \qquad (54)$$

which follow from Eq. (53). Similarly, it is easily checked that all physically valid equations in electrodynamics conserve $\hat{C}$, in the sense that the overall $\hat{C}$ symmetry on both sides is the same. Care must be taken, however, about the fundamental meaning of $\hat{C}$. Its real relevance is in elementary particle physics [52], in which anti-particles which do not occur naturally in the observable universe are manufactured in the laboratory. In the same sense, some chiral chemical compounds such as alkaloids and sugars do not occur in nature (i.e., in the natural universe), but can be synthesized in the laboratory. The absence of these compounds in nature does not imply violation of any discrete symmetry [52]. Similarly, the absence of anti-matter and anti-fields in the natural universe does not imply the violation of $\hat{C}$ symmetry.

2.2 THE $\hat{C}\hat{P}\hat{T}$ THEOREM

That this must be so is embodied in the $\hat{C}\hat{P}\hat{T}$ theorem of field theory [52], of which there are numerous accounts available. The theorem, simply stated, means that the conservation of $\hat{C}\hat{P}\hat{T}$ always appears to occur in the theory of fields and particles. For example:

(1) if $\hat{C}$ is conserved, as we have argued, $\hat{P}\hat{T}$ is conserved. Conversely, if $\hat{C}$ is violated, $\hat{P}\hat{T}$ must be

The $\hat{C}\hat{P}\hat{T}$ Theorem

violated.

(2) If $\hat{P}\hat{T}$ is conserved, then, (a) $\hat{P}$ and $\hat{T}$ are both conserved separately, or, (b) $\hat{P}$ and $\hat{T}$ are both violated.

The discrete symmetry operators $\hat{P}$ and $\hat{T}$ can be applied within electrodynamics with care. Since $\hat{C}$ takes us outside the boundaries of conventional electrodynamics, it follows that $\hat{P}\hat{T}$ *must always be conserved within electrodynamics*. Since $\hat{T}$ violation has been observed experimentally only on one occasion, and then indirectly through $\hat{C}\hat{P}$ violation [52], it is inferred that $\hat{P}$ and $\hat{T}$ are separately conserved in electrodynamics, whose physical equations conserve $\hat{P}$ and $\hat{T}$. Taking the novel Eqs. (4) as an example, the separate conservation of $\hat{P}$ and $\hat{T}$ may be demonstrated as follows. (It is sufficient to demonstrate the conservation of only one of $\hat{P}$ and $\hat{T}$, because it follows that the other must be conserved because $\hat{P}\hat{T}$ is conserved.)

2.2.1 $\hat{P}$ Conservation

Taking Eq. (4a),

$$\boldsymbol{B}^{(1)} \times \boldsymbol{B}^{(2)} = iB^{(0)}\boldsymbol{B}^{(3)}, \tag{55}$$

the left hand side is the cross product of two axial vectors, so the right hand side must also be an axial vector [55]. An axial vector is $\hat{P}$ positive [55] and the scalar $B^{(0)}$ is also $\hat{P}$ positive [10]. Eq. (55) therefore conserves $\hat{P}$, because the overall $\hat{P}$ symmetry on both sides is the same. Similarly, the novel free space equation [9]

$$\boldsymbol{E}^{(1)} \times \boldsymbol{E}^{(2)} = ic^2 B^{(0)}\boldsymbol{B}^{(3)}, \tag{56}$$

conserves $\hat{P}$ because the left hand side is the cross product of two polar vectors, an axial vector $\boldsymbol{B}^{(3)}$.

2.2.2 $\hat{T}$ Conservation

In demonstrating independently $\hat{T}$ conservation in Eqs. (4) it is necessary to account for the rotational motion of the electromagnetic plane wave. By definition, $\hat{T}$ reverses motion, while $\hat{P}$ is not directly concerned with motion. Specifically, the conjugate product is $\hat{T}$ negative. To see this, it is necessary [55] to write out the effect of $\hat{T}$ on the individual fields, for example $\boldsymbol{E}^{(1)}$ and $\boldsymbol{E}^{(2)}$:

$$\hat{T}(\boldsymbol{E}_L^{(1)}) = \boldsymbol{E}_R^{(2)}, \quad \hat{T}(\boldsymbol{B}_L^{(1)}) = -\boldsymbol{B}_R^{(2)}, \quad \hat{T}(\boldsymbol{E}_R^{(1)}) = \boldsymbol{E}_L^{(2)}, \quad \hat{T}(\boldsymbol{B}_R^{(1)}) = -\boldsymbol{B}_L^{(2)} \quad (57)$$

$$\hat{T}(\boldsymbol{E}_L^{(2)}) = \boldsymbol{E}_R^{(1)}, \quad \hat{T}(\boldsymbol{B}_L^{(2)}) = -\boldsymbol{B}_R^{(1)}, \quad \hat{T}(\boldsymbol{E}_R^{(2)}) = \boldsymbol{E}_L^{(1)}, \quad \hat{T}(\boldsymbol{B}_R^{(2)}) = -\boldsymbol{B}_L^{(1)},$$

so:

$$\hat{T}(\boldsymbol{E}^{(1)} \times \boldsymbol{E}^{(2)})_L = (\boldsymbol{E}^{(2)} \times \boldsymbol{E}^{(1)})_R = -(\boldsymbol{E}^{(1)} \times \boldsymbol{E}^{(2)})_R,$$

$$\hat{T}(\boldsymbol{B}^{(1)} \times \boldsymbol{B}^{(2)})_L = (\boldsymbol{B}^{(2)} \times \boldsymbol{B}^{(1)})_R = -(\boldsymbol{B}^{(1)} \times \boldsymbol{B}^{(2)})_R,$$
(58)

where R and L denote right and left circular polarization respectively. The field $\boldsymbol{B}^{(3)}$, being a magnetic field, is also $\hat{T}$ negative [55] and $B^{(0)}$ is $\hat{T}$ positive, being a scalar, so that $\hat{T}$ is conserved.

Similarly, it may be shown [56] that all Eqs. (4) conserve $\hat{T}$, as well as $\hat{P}$ and $\hat{C}$, and therefore conserve all seven discrete symmetries of nature. Furthermore, all physically valid equations of electrodynamics conserve the seven discrete symmetries. There are several phenomena of elementary particle physics [52] in which violation occurs experimentally of one of the discrete symmetries, for example beta decay, but in electromagnetism, it appears that such violation does not occur unless the photon is taking part in some process involving other elementary particles. In free space electromagnetism, it is presumably true that we are dealing only with one type of particle, the photon, and that no symmetry violation occurs in the equations of free space electrodynamics.

2.3 THE CONCEPT OF PHOTON AND ANTI-PHOTON

The concept of anti-particle emerged from the Dirac equation for particles with mass, spin and charge such as the

The Concept of Photon and Anti-Photon

electron. The $\hat{C}$ operator generates the anti-particle from the original particle, a process which is accompanied by the reversal of charge, baryon number, lepton number and so on [52], but which leaves mass and spin unchanged. It is conventionally asserted that the photon is its own anti-photon [52], because the photon is regarded as having no charge, baryon number, lepton number, or strangeness. However, the photon does not exist in isolation of its concomitant fields, embodied in A_μ, and $\hat{C}$ reverses A_μ. Therefore the operator, $\hat{C}$, that generates the anti-particle from the particle also reverses the sign of A_μ. For this reason it is inferred as already described that $\hat{C}$ generates the anti-field from the original field, in the same way as it generates anti-matter from matter. Indeed, in general relativity there is no distinction between field and matter, so that it follows that $\hat{C}$ must generate an anti-field concomitant with the photon defined conventionally as the light quantum $h\nu$.

Therefore it is not sufficient to assert that the photon is its own anti-photon, even though the photon is usually considered to be governed by a different equation of motion (d'Alembert or Proca equation) from a particle such as the electron (Dirac equation). Even this statement which occurs in many standard texts [54] needs qualification, because, following Barut [46], the photon can be described through the same type of equation of motion as the neutrino. This is accomplished through the use of a complex field vector $\boldsymbol{E}/c + i\boldsymbol{B}$, which forms the space-like component of the four-vector G_μ, whose time-like component is zero, $G_\mu = 0$. The Maxwell equations are then [46]

$$\beta_\nu \frac{\partial G_\mu}{\partial x_\nu} = -\frac{j_\mu}{c}, \tag{59}$$

where j_μ is a four-current and where β_ν are 4 x 4 Hermitian matrices definable in terms of Pauli matrices. The four-component equations (59) splits into two two-component equations in a similar manner to the neutrino equation, but unlike it, these two equations are coupled through the subsidiary condition $G_0 = 0$. Equation (59) has the general form [46] of the wave equation for all spin values, and the Dirac equation [46] *has the same form*. So do the Kemmer equations, which are related [57] to the Proca equation for a photon of finite mass.

In view of these results, which are not well known, but

nevertheless rigorous [46] in special relativity, it is inferred that there is an anti-photon, whose mass and boson spin are the same as those of the photon, but whose concomitant anti-field is reversed in sign. The photon can therefore be defined as $h\nu$ accompanied by A_μ, whose amplitude is $A^{(0)}$; and the anti-photon by $h\nu$ accompanied by $-A_\mu$, whose amplitude is $-A^{(0)}$. As discussed already, however, the anti-photon appears not to exist in nature, but can be manufactured in the laboratory, for example from a radiating positron. There is no doubt that the $\hat{C}$ parity of the photon when designated by the standard symbol γ in elementary particle physics is negative,

$$\hat{C}(\gamma) = -\gamma, \qquad (60)$$

and is used as such in standard texts [52]. In the manufacture of laser radiation, however, and in similar laboratory phenomena, electromagnetism is generated from electrons, either free or bound within atoms, and not from positrons. The sign of $A^{(0)}$ in laser radiation is therefore fixed as we have seen, and in particular, the sign of the scalar amplitude $B^{(0)}$ of $\boldsymbol{B}^{(1)}$, $\boldsymbol{B}^{(2)}$ and $\boldsymbol{B}^{(3)}$ is positive definite. The sign of $\boldsymbol{B}^{(3)}$ from a radiating positron is expected to be opposite from that of $\boldsymbol{B}^{(3)}$ from a radiating electron, and this would be a test for the anti-photon.

2.4 PARITY OF THE PHOTON AND ANTI-PHOTON

A text on elementary particles and symmetries such as that of Ryder [52] states that the photon is described by A_μ, and under the parity inversion operator A_μ behaves as follows,

$$A_\mu := \epsilon_0\left(\mathbf{A}, \frac{i\phi}{c}\right) \xrightarrow{\hat{P}} \epsilon_0\left(-\mathbf{A}, \frac{i\phi}{c}\right), \qquad (61)$$

and so it is stated that the photon has negative *intrinsic* parity. The latter is well defined in elementary particle physics [52] only for uncharged and non-strange bosons, particles which can be created singly. This statement is a consequence of the law of conservation of charge, and since neither the photon nor the anti-photon as defined in Sec. 2.3 is charged, it is possible for both to have a well defined negative intrinsic parity. The equivalent of Eq. (61) for the anti-photon is

$$-A_\mu := -\epsilon_0\left(\mathbf{A}, \frac{i\phi}{c}\right) \xrightarrow{\hat{P}} \epsilon_0\left(\mathbf{A}, -\frac{i\phi}{c}\right), \qquad (62)$$

because the photon and anti-photon are identical in our definition except for the sign of $A^{(0)}$. Since $A^{(0)}$ is not spatio-temporal, and parity is by definition spatio-temporal, the intrinsic parity of photon and anti-photon must be the same, i.e., negative. Note that if the photon is defined as the light quantum of energy $h\nu$, a scalar, then its intrinsic parity cannot be deduced. For this purpose, the classical quantity A_μ is assumed to be sufficient, even in the context of quantized elementary particle theory [52]. This shows clearly that the "photon" is not fully defined in the absence of concomitant fields. The fact that photon and anti-photon have the same intrinsic parity is consistent with the theorem [52] that for bosons, particle and anti-particle must have the same parity. (For fermions [52], particle and anti-particle have opposite parity.)

2.5 MOTION REVERSAL SYMMETRY OF PHOTON AND ANTI-PHOTON

The motion reversal symmetry ($\hat{T}$) of the photon is expressed [52] again in terms of the potential four-vector A_μ, i.e., through the concomitant field,

$$A_\mu := \epsilon_0\left(\mathbf{A}, \frac{i\phi}{c}\right) \to \epsilon_0\left(-\mathbf{A}, \frac{i\phi}{c}\right), \qquad (63)$$

and is negative. Similarly, the $\hat{T}$ symmetry of the anti-photon, defined in this way, is also negative. However, it is obviously necessary to be clear about which photon property is being referred to in this definition. The photon linear momentum whose magnitude is $h\nu/c$, is negative to $\hat{T}$. The photon angular momentum whose magnitude is $\hbar$, is also negative to $\hat{T}$. However, if the photon is defined as the unit of energy $h\nu$, its motion reversal symmetry is positive by definition. These motion reversal symmetries are the same for the anti-photon, whose energy and linear and angular momenta are defined as positive to $\hat{C}$. Since physical photon energy is always proportional to the square of the concomitant electric or magnetic field, its energy is always positive. Presumably, the Kemmer equation (59), which is structured identically with the Dirac equation, also produces the unphysical negative energies [54] that originally led Dirac to propose the concept of anti-particle. Thus, the Kemmer equation would lead to the anti-photon by analogy.

2.6 THE PHOTON'S CHARGE CONJUGATION SYMMETRY, $\hat{C}$

It has already been inferred that the $\hat{C}$ symmetry of the photon is negative, an inference once more obtained through the concomitant field, but this time through the change in sign of the *scalar* magnitude $A^{(0)}$, and not by definition of $\hat{C}$ [52], through any spatio-temporal quantity. The photon is an eigenstate, ζ, of $\hat{C}$ such that

$$\hat{C}\zeta = -\zeta, \tag{64a}$$

and so has *negative charge parity* ϵ_c. This result is denoted by

$$\hat{C}\zeta = \epsilon_c \zeta, \quad \epsilon_c = -1, \tag{64b}$$

in elementary particle physics, which, however, asserts that the photon is its own anti-photon because it carries no charge, lepton number, baryon number or strangeness. We have argued, however, that since the photon always has its concomitant A_μ, it is possible to define the anti-photon through the change in sign of $A^{(0)}$ upon application of $\hat{C}$.

Clearly, a more consistent scheme is needed in view of the recent discovery of $\boldsymbol{B}^{(3)}$ through equations (4), because the direction of $\boldsymbol{B}^{(3)}$, which is in principle measurable in the laboratory, is opposite for photon and anti-photon, the former being produced by a radiating electron and the latter by a radiating positron. One possible scheme is suggested as follows.

2.7 SCHEME FOR THE PHOTON'S FUNDAMENTAL SYMMETRIES, $\hat{C}$, $\hat{P}$, AND $\hat{T}$

Equations (4), defining the novel spin field $\boldsymbol{B}^{(3)}$ of electrodynamics [9-15] are invariant to $\hat{C}$, $\hat{P}$, and $\hat{T}$ for the photon and anti-photon. A set of self consistent rules is therefore necessary by which the symmetries of the electromagnetic fields in free space ("in vacuo") can be classified without ambiguity. For this purpose it is inferred that two appropriate variables are the helicity, a scalar denoted λ, and A_μ. The photon is therefore identified as (λ, A_μ), and the anti-photon as $(\lambda, -A_\mu)$, where $A^{(0)}$ is positive definite and negative definite respectively. This definition allows for wave - particle dualism, in that the photon is both a particle, and an electrodynamic wave. It is now clear that the particle generates the spin $\pm \hbar$, and the wave generates the *spin* fields $\pm B^{(0)} \hbar$. The existence of the spin field is an inevitable consequence of the existence of $\hbar$, the spin angular momentum of the photon, and the fundamental Dirac constant.

Care is needed in the definition of helicity. In the classical relativistic theory of electromagnetism, considered to be without concomitant mass, the helicity of the photon is +1 or -1, and emerges following the work of Wigner, as described for example by Ryder [54], from considerations of the Pauli-Lubansky vector and the generator of space-time translations. The numerical values of the helicity are therefore basic symmetry properties of the Poincaré group and, anticipating the particle interpretation in quantum field theory, are (within to a factor $\hbar$) the eigenvalues of spin angular momentum for the boson with a missing zero. This boson is of course the massless photon. It is important to realize that if the photon has mass then the eigenvalues are $-\hbar$, 0, and $\hbar$, and the obscure little group [54] E(2) introduced by Wigner is replaced by the physically meaningful rotation group. This is one of the major theoretical advantages of regarding the photon as massive, rather than massless [9–15]. There is, strictly speaking, no non-relativistic counterpart of the Wigner definition of helicity, but the working definition [52],

$$\lambda = \sigma \cdot \frac{p}{|p|}, \qquad (65)$$

is often used. Here σ denotes spin, and p denotes linear momentum. Therefore the helicity is loosely defined as the component of spin along the direction of motion, and is therefore a pseudo-scalar negative to $\hat{P}$. From Eq. (11) it is seen that κ is in a sense also a pseudo-scalar, negative to $\hat{P}$, and indeed, the ratio of the *magnitude* of photon intrinsic angular momentum, i.e., $\hbar$, to the photon linear momentum magnitude, $\hbar\omega/c$, is $1/\kappa$, which has the dimensions of wavelength, also given the symbol λ. The angular momentum magnitude of the photon switches sign with circular polarization and so does the helicity. The latter embodies the result that electromagnetism in free space is a phenomenon involving simultaneous rotation and translation, and the photon as particle has both linear and angular momentum. These momenta are intrinsically and inevitably relativistic if there is no mass, and the massless photon is never at rest. The helicity summarizes these spatio-temporal properties in just two numbers, -1 and +1, or if there is mass, in three numbers, -1, 0, and +1. The electromagnetic character concomitant with the helicity is A_μ. The description of the photon as (λ, A_μ) is therefore fully relativistic as required, bearing in mind that in the relativistic quantum field theory $\hat{A}_\mu$ is an operator [54], whose expectation value is A_μ. In this relativistic definition λ is defined by the rigorous Wigner method, strictly speaking classical in nature [54].

The effect of $\hat{C}$, $\hat{P}$, and $\hat{T}$ on the photon can now be defined as

$$(\lambda, A_\mu, A^{(0)}) \xrightarrow{\hat{C}} (\lambda, -A_\mu, -A^{(0)}), \tag{66a}$$

$$(\lambda, \mathbf{A}, \phi) \xrightarrow{\hat{P}} (-\lambda, -\mathbf{A}, \phi), \tag{66b}$$

$$(\lambda, \mathbf{A}, \phi) \xrightarrow{\hat{T}} (\lambda, -\mathbf{A}, \phi), \tag{66c}$$

$$A_\mu := \epsilon_0 \left(\mathbf{A}, \frac{\phi}{c} \right), \tag{66d}$$

which are also the symmetry properties of the classical electromagnetic wave in vacuo. Therefore, the $\hat{C}$ operator leaves the space-time quantity λ unchanged by definition, while changing the sign of $A^{(0)}$ by definition. $\hat{C}$ thus produces a distinct entity which we identify classically as the anti-wave and quantum mechanically as the anti-photon. The photon is its own anti-photon [52] only in regard to the spatio-temporal helicity, λ, in Eqs. (66).

The $\hat{P}$ operator reverses the sign of λ by definition. The $\hat{P}$ symmetry of the space-like part of A_μ (the vector potential) is negative, and that of the time-like part (the scalar potential ϕ) is positive. $\hat{P}$ produces the classical wave or quantized photon with opposite helicity. The $\hat{T}$ operator does not change the sign of λ, and the $\hat{T}$ symmetry of the space-like part of A_μ is negative, while that of the

Scheme for the Photon's Fundamental Symmetries

time-like part is positive.

In Eqs. (66) it is assumed implicitly that the scalar potential ϕ is non-zero, as in the Lorentz gauge defined by the condition [43]

$$\frac{\partial A_\mu}{\partial x_\mu} = 0, \qquad (67)$$

in Minkowski notation $x_\mu := (X, Y, Z, ict)$. In the transverse gauge, the scalar potential ϕ is zero, but this makes no difference to the scheme described by Eq. (66). In summary therefore, the conventional inference [52] that the photon is its own anti-photon cannot be true if account is taken of the concomitant fields, described by A_μ, and the photon is always concomitant with $\hat{C}$ negative electric and magnetic fields.

2.8 SYMMETRY OF THE PURE IMAGINARY FIELD $i\boldsymbol{E}^{(3)}/c$

In Chap. 1, and in dealing in Sec. 2.3 with the Kemmer Eq. (59), use was made of the imaginary fields $i\boldsymbol{E}/c$ and $i\boldsymbol{B}$, together with the cyclically symmetric relation

$$\boldsymbol{E}^{(1)} \times \boldsymbol{E}^{(2)} = -E^{(0)}(i\boldsymbol{E}^{(3)})^*, \qquad (68)$$

(with cyclic permutations of (1), (2) and (3)). The fields being permuted here are $\boldsymbol{E}^{(1)}$, $\boldsymbol{E}^{(2)}$ and $i\boldsymbol{E}^{(3)}$; and not $\boldsymbol{B}^{(1)}$, $\boldsymbol{B}^{(2)}$ and $\boldsymbol{B}^{(3)}$ as in Eq. (4). Thus $\boldsymbol{B}^{(3)}$ is pure real in Eq. (4) and $i\boldsymbol{E}^{(3)}$ pure imaginary in Eq. (68). (The negative sign on the right hand side of the latter equation is due to $i\boldsymbol{E}^{(3)} = -(i\boldsymbol{E}^{(3)})^*$ where * denotes "complex conjugate" as usual.) Using $E^{(0)} = cB^{(0)}$ in free space and comparing Eqs. (4) and (68), it is seen that $i\boldsymbol{E}^{(3)}/c$ must have the same discrete symmetries as $\boldsymbol{B}^{(3)}$. This is confirmed through the use of the sum $\boldsymbol{E}/c + i\boldsymbol{B}$ in the Kemmer equation (59).

Another indication that $\boldsymbol{B}^{(3)}$ and $-i\boldsymbol{E}^{(3)}/c$ have the same discrete symmetries, $\hat{C}$, $\hat{P}$, and $\hat{T}$, comes from the duality transformation of special relativity [46], under which the Maxwell equations are invariant in vacuo. In S.I. units, the duality transformation is equivalent to the space-like transformations:

$$\boldsymbol{B} \rightarrow -\frac{i\boldsymbol{E}}{c}, \quad \boldsymbol{E} \rightarrow ic\boldsymbol{B}, \quad \boldsymbol{B}^* \rightarrow i\boldsymbol{E}^*/c, \quad \boldsymbol{E}^* \rightarrow -ic\boldsymbol{B}^*, \qquad (69)$$

Chapter 2. Fundamental Symmetries

the change in sign being occasioned again by the fact that i changes sign under complex conjugation. The factor i appears in the duality transformation because of the pseudo-Euclidean (i.e., complex) nature of Minkowski space-time, whose space-like part is real, and whose time-like part is imaginary. For conventional Maxwellian *wave* fields in free space, it is easy to see that the duality transformation simply generates the complex electric wave from the complex magnetic wave. For example, for the conjugate wave field with circular coordinate (1) we have

$$\boldsymbol{B}^{(1)} = \frac{B^{(0)}}{\sqrt{2}}(i\boldsymbol{i} + \boldsymbol{j})e^{i\phi} \rightarrow \frac{iE^{(0)}}{\sqrt{2}c}(\boldsymbol{i} - i\boldsymbol{j})e^{i\phi} = \frac{i\boldsymbol{E}^{(1)}}{c}, \qquad (70)$$

and for the complex conjugate, circular coordinate (2), we have

$$\boldsymbol{B}^{(2)} = \frac{B^{(0)}}{\sqrt{2}}(-i\boldsymbol{i} + \boldsymbol{j})e^{-i\phi} \rightarrow \frac{-iE^{(0)}}{\sqrt{2}c}(\boldsymbol{i} + i\boldsymbol{j})e^{-i\phi} = \frac{-i\boldsymbol{E}^{(2)}}{c}. \qquad (71)$$

The wave fields are complex in nature, but the *spin* field $\boldsymbol{B}^{(3)}$ is *pure real* [9–15]. Therefore the duality transformation applied to $\boldsymbol{B}^{(3)}$ must generate a *pure imaginary* electric field $-i\boldsymbol{E}^{(3)}/c$ because Minkowski space-time is complex. This is the fundamental reason in relativity why there cannot be a longitudinal real $\boldsymbol{E}^{(3)}$, and as we have seen, electromagnetism in free space has no non-relativistic limit, the free space Maxwell equations are invariant under a Lorentz transformation [4] but not under the Galilean transformation. This inference, as is well known [4], led to the development of the theory of special relativity by Lorentz, Poincaré and Einstein. Following the usual rule that pure real fields are physical [4, 45], pure imaginary fields are unphysical, it follows that $-i\boldsymbol{E}^{(3)}/c$ has no physical significance at first order, i.e., $i\boldsymbol{E}^{(3)}$ is not a physically significant electric field, meaning that it does not act on matter as an electric field. In contrast, $\boldsymbol{B}^{(3)}$ is a pure real and therefore physically significant magnetic field, as evidenced, for example, through Eq. (6), defining the inverse Faraday effect at second order. (There is also (Chap. 12) a first order inverse Faraday effect in the physical magnetic field $\boldsymbol{B}^{(3)}$ [9–15].) No electric analogue of the inverse Faraday effect appears to have been reported experimentally, from which it is consistent to infer that $i\boldsymbol{E}^{(3)}$ is unphysical.

Symmetry of the Pure Imaginary Field $i\mathbf{E}^{(3)}/c$

This conclusion is also consistent with the fact that the discrete symmetries $\hat{P}$ and $\hat{T}$ of physical magnetic and electric fields are different. The former is axial, $\hat{P}$ positive, $\hat{T}$ negative, the latter is polar, $\hat{P}$ negative, $\hat{T}$ positive [9–15]. It is not possible, by fundamental geometry, to produce a polar vector in Euclidean space from the vector cross product of two polar or two axial vectors, meaning that the pure imaginary $i\mathbf{E}^{(3)}$ in Eq. (68) must be regarded as *axial* i.e., as having the same symmetry as a pure real magnetic field. This is consistent with the fact that the duality transformation converts a real, axial, and physical magnetic field into an imaginary electric field which is unphysical at first order. Seen in another way, the conjugate product, as inferred from Eq. (68), is $\hat{T}$ *negative* and cannot generate a real, physical, phase free "electric spin field". This is consistent, finally, with the fact that the real magnetic field $\mathbf{B}^{(3)}$ is directly proportional to an angular momentum, which is $\hat{T}$ negative, $\hat{P}$ positive.

2.9 EXPERIMENTAL DEMONSTRATION OF THE EXISTENCE OF THE ANTI-PHOTON

The existence of the anti-photon can be demonstrated experimentally simply by a re-examination of the various phenomena of magnetization by light, such as the inverse Faraday effect [19–26]; the optical Faraday effect [25]; light shifts in atomic spectra [58, 59]; and recent novel manifestations such as optically induced nuclear magnetic resonance shifts [60–62]. The reason is that these effects, which are field-matter phenomena, can be interpreted in terms of Eq. (4), in which the conjugate product is replaced by a term involving $\mathbf{B}^{(3)}$ at second order in $B^{(0)}$. They are therefore experimental demonstrations of the existence of $\mathbf{B}^{(3)}$. For example the inverse Faraday effect can be interpreted directly in terms of $\mathbf{B}^{(1)} \times \mathbf{B}^{(2)}$ and therefore of $iB^{(0)}\mathbf{B}^{(3)}$ as demonstrated recently by Woźniak *et al.* [63].

Application of $\hat{C}$ to these experiments shows that there must be corresponding anti-field / anti-matter experiments which demonstrate the existence of $-\mathbf{B}^{(3)}$, i.e., of $\mathbf{B}^{(3)}$ with the sign of $B^{(0)}$ reversed. It is possible to infer on these experimental grounds that there is an anti-photon whose concomitant $B^{(0)}$ is reversed in sign.

A direct demonstration of the existence of the anti-photon requires direct experimental proof that the sign of

$B^{(3)}$ (and therefore of $B^{(0)}$) is reversed under appropriate circumstances. For example, in one of the simplest cases, the sign of $B^{(3)}$ should be reversed in circularly polarized radiation from a moving positron when compared with radiation from a moving electron. Unequivocal proof in such an experiment would require the direct observation of $B^{(3)}$ at first order, because at second order, the sign of $B^{(0)2}$ is unchanged (i.e., the squares of $B^{(0)}$ and $-B^{(0)}$ are both positive). It is emphasized that the source of radiation must be *anti-matter*, it is obviously insufficient simply to replace a radiating anion by a radiating cation, because both are made of matter (electrons, protons, neutrons etc., and not positrons, anti-protons, and anti-neutrons). Radiation from the cation and anion (or *any* type of microscopic or macroscopic matter) always has a positive definite scalar magnitude $B^{(0)}$ because the source always has a negative definite e. Radiation from anti-matter has a negative definite $B^{(0)}$ because the source has a positive definite e. It may be possible to infer under certain circumstances in elementary particle physics the participation of the anti-photon in reactions among, or scattering of, elementary particles [52].

CHAPTER 3. THE ORIGINS OF WAVE MECHANICS

Wave mechanics is among the foremost achievements of natural philosophy, and there are numerous accounts of its inception, notable among which is the scholarly monograph by Pais [64]. The old quantum theory was based on the ideas of Planck, Einstein, and others in the first decade of this century, ideas centered on the light quantum hypothesis as described by Pais [64] by careful reference to the original papers, warts and all. The light quantum could not have arisen without the photon, which is therefore the fundamental idea of the old quantum theory, a theory based essentially on the consideration of electromagnetism in thermodynamic equilibrium with matter, i.e., on a field-matter hypothesis. As described briefly in Chap. 1, the recent realization that there exists a spin field $\boldsymbol{B}^{(3)}$ in free space makes no difference to these original ideas, i.e., $\boldsymbol{B}^{(3)}$ is consistent with the ideas of wave mechanics, both for the photon without mass, and the photon with mass. The emphasis on $\boldsymbol{B}^{(3)}$ in this chapter is a convenient way of demonstrating the origins of wave mechanics, making it clear as a by-product of the analysis that the spin field $\boldsymbol{B}^{(3)}$ is precisely defined as *an expectation value*, or stationary state, of the electromagnetic wave fields in free space.

The transition from the old quantum theory to wave mechanics took place in a series of profound philosophical developments from about 1900 to about 1925 as the result of the combined efforts of many thinkers. It is generally accepted that one of the key inferences was that of Louis de Broglie, whose Doctoral Thesis proposed the idea of matter waves [64, 65]. As in Chap. 1, these are essentially electromagnetic waves with finite rest mass, from which it is natural to propose that the photon itself may have mass. This line of thought (initiated in about 1916 by Einstein in the context of classical general relativity [66]) was pursued by de Broglie in a series of monographs [67] and scientific papers, one of the by products of which was the proposal of the Proca equation in 1930 [68], essentially the d'Alembert equation with mass, a wave equation. From matter waves to wave mechanics is in some ways a short step mathematically, but based on de Broglie's giant conceptual leap. In about 1925 Schrödinger proposed an equation [69] which was based on

Chapter 3. The Origin of Wave Mechanics

a novel psi function, Ψ, which is operated upon by a Hamiltonian operator $\hat{H}$ to generate a scalar energy eigenvalue En. The psi function is therefore an eigenfunction, and the time-independent Schrödinger equation an eigenvalue equation. It was proposed by Schrödinger that his equation

$$\hat{H}\Psi = En\Psi, \qquad (72)$$

be an equation of motion that applies to matter, in which, following de Broglie, there exist matter *waves*. Despite the fact that the Schrödinger equation in this form is not a second order differential wave equation, it was derived in direct response to the wave nature of light and the inferred wave nature of matter. Essentially speaking, fields and matter are both undulatory and simultaneously particulate.

It is well known [64] that the physical nature of the psi function was not immediately apparent, and that it was left to Born in about 1926 [64, 70] to propose an interpretation based on probability, the Copenhagen interpretation, but one which was never accepted by Einstein, who continued to refer to Ψ simply as the psi function [64]. In the Copenhagen interpretation of light and matter, waves and particles never co-exist [71, 72], but in the Einstein-de Broglie interpretation, they can. The latter interpretation has recently received renewed experimental support in double slit experiments which demonstrate the simultaneous presence of single photons and waves [72].

The interpretation of the wave function by Born was itself based on an analogy with the nature of the electromagnetic phase, ϕ, used in Chap. 1, and on the fact that electromagnetic energy is quadratic in the electric and magnetic fields. The identification [70] of ϕ with *action* S:

$$S = \hbar\phi = \hbar\kappa^\mu x_\mu, \qquad (73)$$

is the starting point of our interpretation of the newly discovered spin field $\boldsymbol{B}^{(3)}$ [9-15], which does not occur in standard electrodynamics [4]. This method rigorously defines $\boldsymbol{B}^{(3)}$ as a stationary state of the electromagnetic field in vacuo. In so doing, the close relationship between wave mechanics (or quantum mechanics) and the particle-wave nature of light and matter emerges naturally.

3.1 THE PHASE AS WAVE FUNCTION

Our account here is based on that of Atkins [70], but is centered on the emergence from classical electrodynamics [9–15] of the spin field $B^{(3)}$ and on the consequent need to interpret it within quantum field theory. Since action in classical mechanics has the units of energy multiplied by time, it is clear that

$$S = \hbar(\omega t - \boldsymbol{\kappa} \cdot \boldsymbol{r}) = (\hbar \omega) t - \hbar \boldsymbol{\kappa} \cdot \boldsymbol{r}, \tag{74}$$

has the correct units. The analogy between electromagnetic phase in radiation and action in matter is the key to the derivation of the time-dependent Schrödinger equation as described by Atkins [70]. This is also the key to the interpretation of $B^{(3)}$ as an expectation value of the electromagnetic field between complex conjugate eigen-functions which will be identified, for one photon, as the electromagnetic phase $e^{i\phi}$, associated with the circular coordinate (1), and the phase $e^{-i\phi}$, identified with circular coordinate (2). It follows that the classical $B^{(3)}$ becomes an operator $\hat{B}^{(3)}$ which is described by the Schrödinger equation for one photon,

$$\hat{B}^{(3)} |\Psi\rangle = \pm B^{(0)} |\Psi\rangle, \tag{75}$$

and is therefore precisely defined in wave mechanics.

The time dependent Schrödinger equation is essentially [70] the hypothesis that a particle is described by a wave like relation between amplitudes Ψ at different points in phase space:

$$\Psi(x_2, t_2) = e^{iS/\hbar} \Psi(x_1, t_1). \tag{76}$$

The particle is therefore associated with a wave-like amplitude Ψ, and propagates along a path that makes $\phi = S/\hbar$ a minimum. The scalar ϕ is identified with action as in Eq. (73), thus making a direct link between Fermat's principle of least time in optics and Hamilton's principle of least action in mechanics. This link is the essence of wave mechanics, because wave and particle properties are identified through Eq. (73). The time dependent Schrödinger equation is essentially Eq. (76), where S is the action associated with the trajectory from (x_1, t_1) to (x_2, t_2). Equation (76) can be rearranged without further physical insight using the relation between action and energy,

$$En = -\frac{\partial S}{\partial t}, \qquad (77)$$

and using the time-independent Schrödinger equation (72), and is conventionally written as

$$\hat{H}\Psi = i\hbar \frac{\partial \Psi}{\partial t}. \qquad (78)$$

The identification of electromagnetic phase with mechanical action, Eq. (73), is therefore the basis of wave (or quantum) mechanics.

3.2 THE WAVE MECHANICS OF A SINGLE PHOTON IN FREE SPACE

The classical electromagnetic exponents for monochromatic radiation at the angular frequency ω are

$$\Psi = e^{-i(\omega t - \kappa \cdot r)} = \psi(r) e^{-i\omega t}, \qquad \Psi^* = e^{i(\omega t - \kappa \cdot r)} = \psi^*(r) e^{i\omega t}, \qquad (79)$$

where * denotes "complex conjugate" as usual. That these are *wave functions for the single photon* becomes clear from the fact that Eqs. (79) are solutions of the time dependent Schrödinger equation written in the form

$$\frac{\partial \Psi}{\partial t} = -\frac{i}{\hbar} En\, \Psi, \qquad (80)$$

where the energy is that of one photon, the light quantum $En = \hbar\omega$. The time dependent wave function of a single photon in free space is therefore Ψ. Furthermore, by writing Eq. (80) in the form,

$$i\hbar \frac{\partial}{\partial t} \Psi = En\, \Psi, \qquad (81)$$

it becomes Eq. (72), allowing the identification of *the Hamiltonian operator for a single photon*,

Wave Mechanics of a Single Photon in Free Space

$$\hat{H} := i\hbar\frac{\partial}{\partial t}, \qquad (82)$$

which is a differential operator, operating on the time dependent Ψ function, the classical exponent of the electromagnetic wave. The operator $\hat{H}$ and the energy En are entirely kinetic in free space. Equation (81) is therefore a Schrödinger equation for one photon propagating in free space. Multiplying both sides of Eq. (81) by the complex conjugate Ψ^* gives Eq. (20) of Chap. 1, which, together with the classical electromagnetic equation (19) gives the de Broglie Guiding theorem, Eq. (1), as described in that chapter.

In multiplying Eq. (81) on both sides by the conjugate Ψ^* we have formed the *expectation value* of the one photon Hamiltonian operator in free space. This expectation value is the light quantum $\hbar\omega$, which is related to the *square* of the concomitant electric and magnetic fields of the photon through equations such as (2), (3), and (8) of Chap. 1. This inference in turn leads us to the essence of Born's interpretation of the psi function. This interpretation rests on the axiom that $\Psi^*(\mathbf{r}, t)\Psi(\mathbf{r}, t)\,dV$ is the probability of the particle being in the infinitesimal volume dV at the point r at time t. For one photon in free space this product is dV, and although the individual exponents, which are identified as single photon wave-functions, vary with time, the *conjugate product, in the language of quantum mechanics, is a stationary state of the photon.*

Following Atkins [70], this inference is consistent with the fact that Born was led to his interpretation by Einstein's correlation of the number of photons in a light beam with its intensity, the latter being proportional to *conjugate products* such as integrals over $\mathbf{E}^{(i)}\cdot\mathbf{E}^{(i)*}$ or $\mathbf{B}^{(i)}\cdot\mathbf{B}^{(i)*}$ where (i) runs from (1) to (3) following the discovery of the spin field $\mathbf{B}^{(3)}$, and $*$ denotes "complex conjugate". If the light beam is composed of one photon the psi function is a complex exponential, identified with the phase of the classical electromagnetic wave. The time dependent form of the wavefunction of one photon propagating in free space is therefore $\psi\exp(-i\omega t)$, and the time dependence is [70] a modulation of the phase of the wavefunction: $\exp(-i\omega t)$ oscillates periodically from 1 to $-i$ to i and back to 1 with a frequency ω and a period $1/(2\pi\omega)$, as described diagrammatically by Atkins [70].

The wave mechanics of a single photon in free space can be described equivalently by d'Alembert's equation and

Schrödinger's equation. The Schrödinger equation in this case is therefore rigorously consistent with special relativity, because the d'Alembert equation is already a relativistically correct description. The contemporary understanding of this result can be summarized through the fact that the Klein-Gordon, d'Alembert and Dirac equations of motion for a particle with mass $m = 0$ can be written as

$$\Box \phi_s = \Box A^\mu = \Box \psi_s = 0, \tag{83}$$

where ϕ_s is a scalar field, A_μ is the electromagnetic gauge field (a vector field) and ψ_s is a spinor field. The Klein-Gordon equation is the relativistically correct form of the Schrödinger equation [54] for a particle with non-zero mass, and so for $m = 0$, the relativistically correct form of the Schrödinger equation is given in (83). For one massless photon travelling in free space Eq. (83) shows finally that there is a link between the scalar, vector and spinor fields of the photon, a result which is consistent with the Kemmer equation, the Maxwell equations in free space in neutrino form, as described by Barut [46].

The dynamics of one photon in free space represents a meeting place, or confluence, of several major concepts in classical, quantum and relativistic field theory, and considerations of the phase of the electromagnetic plane wave lead to the wavefunction of quantum mechanics. For one photon, the latter has no statistical nature, but for many photons the contemporary description [73, 74] relies on creation and annihilation operators. Phenomena of light squeezing [74] show that photon statistics are essentially quantum mechanical in nature, but the interpretation of quantum mechanics itself is more than ever a matter for lively debate [75–77]. Recent experiments appear to indicate [75] that the particulate photon and its concomitant wave coexist, thus indicating support for the Einstein-de Broglie interpretation of light. More will be said about this view in later chapters of this book.

3.3 STATIONARY STATES OF ONE PHOTON IN FREE SPACE

The light quantum $\hbar\omega$ is the stationary state defined by the expectation value of the Hamiltonian $\hat{H}$ for one photon, a result summarized by the following set of equations, where the Dirac bracket notation is used,

Stationary States of One Photon in Free Space

$$\langle \Psi^* | \hat{H} | \Psi \rangle = \int \Psi^* \hat{H} \Psi \, dV. \tag{84}$$

Thus,

$$\hbar\omega = \langle \Psi^* | i\hbar \frac{\partial}{\partial t} | \Psi \rangle = \epsilon_0 \int E^{(0)2} dV, \tag{85}$$

is the expectation value from the Schrödinger equation of motion of the photon in free space,

$$\hat{H}|\Psi\rangle = \hbar\omega|\Psi\rangle, \tag{86}$$

where the wavefunction is defined by

$$|\Psi\rangle = e^{-i\phi}, \quad \phi = \omega t - \boldsymbol{\kappa} \cdot \boldsymbol{r}. \tag{87}$$

When there are many photons in the light beam, it is well known that on average, the light beam intensity, I_0, in watts per square meter is given by

$$I_0 = \epsilon_0 c E^{(0)2}, \tag{88}$$

which is therefore an expectation value for many photons, as opposed to just one photon. In calculating I_0 quantum mechanically to give the final result (88), Bose-Einstein statistics may be used. The light quantum $\hbar\omega$ is the essential building block of these statistics, which consider radiation in thermodynamic equilibrium with a black body. In deriving the Planck law using Bose-Einstein statistics, however, [70], the Schrödinger equation differs from Eq. (86), which is that for *one photon in free space*, in which there is no interaction with matter. In other words Eq. (86) is that for the trajectory of one photon in free space. In this case, the wave functions are complex exponentials defined by the phase of the classical electromagnetic plane wave in free space.

Equation (85) represents the stationary state of photon energy in free space. The corresponding stationary states of the photon's linear and angular momenta are given by the Schrödinger equations,

Chapter 3. The Origin of Wave Mechanics

$$\hat{p}|\Psi\rangle = \frac{\hbar\omega}{c}|\Psi\rangle, \quad (89a)$$

$$\hat{J}|\Psi\rangle = \pm\hbar|\Psi\rangle, \quad (89b)$$

where the linear and angular momentum one photon operators are

$$\hat{p} = \frac{i\hbar}{c}\frac{\partial}{\partial t}, \quad \hat{J} = \frac{i\hbar}{\omega}\frac{\partial}{\partial t}. \quad (90)$$

As described in Chap. 1, the angular momentum operator $\hat{J}$ defines the novel spin field $\hat{B}^{(3)}$ in operator form, and in the remainder of this section we show that $\hat{B}^{(3)}$ is a fundamental photon property, the longitudinal photomagneton [9-15].

In quantum mechanics, independently [70] of any consideration of the Schrödinger equation of motion, the angular momentum is associated with a projection (or azimuthal) quantum number M_J. For the photon without mass this implies that

$$\hat{J}|\Psi\rangle = M_J\hbar|\Psi\rangle = \pm\hbar|\Psi\rangle, \quad (91)$$

an equation which can be rewritten in terms of the *rotation generator* $\hat{J}^{(3)}$, an operator whose expectation value [14, 15] is the unit axial vector **k** in the direction of photon propagation, Z:

$$\frac{\hat{J}}{\hbar}|\Psi\rangle := \hat{J}^{(3)}|\Psi\rangle = \pm 1|\Psi\rangle. \quad (92)$$

This gives the well known result [54, 70] that rotation generators are angular momentum operators of quantum mechanics within a factor $\hbar$. Since Eq. (92) for the photon in free space must be rigorously relativistic, the 4 x 4 rotation generator must be used [54]. In matrix form,

$$\hat{J}^{(3)} = \begin{bmatrix} 0 & -i & 0 & 0 \\ i & 0 & 0 & 0 \\ 0 & 0 & 0 & 0 \\ 0 & 0 & 0 & 0 \end{bmatrix} \left(= \begin{bmatrix} 0 & -i & 0 \\ i & 0 & 0 \\ 0 & 0 & 0 \end{bmatrix} \right), \quad (93)$$

where the 3 x 3 Euclidean form is given in brackets for

reference. Therefore the rotation generator $\hat{J}^{(3)}$ is an operator, which operates on the wave function $|\Psi\rangle$. In differential form it is, $\hat{J}^{(3)} = (i/\omega)(\partial/\partial t)$. The matrix and differential forms of $\hat{J}^{(3)}$ are equivalent [54]. The expectation value of $\hat{J}^{(3)}$ is, in Dirac notation, and analogously with Eq. (85),

$$\boldsymbol{k} = \langle \Psi^* | \hat{J}^{(3)} | \Psi \rangle, \qquad (94)$$

where $\boldsymbol{k}$ is our axial unit vector in Z. However, Eq. (4a) of Chap. 1 shows that the same axial unit vector is proportional to the conjugate product, $\boldsymbol{B}^{(1)} \times \boldsymbol{B}^{(2)}$ which *experimentally* mediates the inverse Faraday effect, and which is therefore a physical property of light:

$$\boldsymbol{k} = -\frac{i}{B^{(0)2}} \boldsymbol{B}^{(1)} \times \boldsymbol{B}^{(2)} = \frac{\boldsymbol{B}^{(3)}}{B^{(0)}} = \langle \Psi^* | \frac{\hat{J}}{\hbar} | \Psi \rangle. \qquad (95)$$

This equation leads directly to the definition of the photomagneton in terms of the angular momentum operator of one photon in free space:

$$\boldsymbol{B}^{(3)} = \langle \Psi^* | \hat{B}^{(3)} | \Psi \rangle, \qquad \hat{B}^{(3)} = B^{(0)} \frac{\hat{J}}{\hbar}. \qquad (96)$$

The latter can also be obtained by multiplying both sides of the defining Schrödinger Eq. (92) by $B^{(0)}$, the magnetic flux density amplitude in free space of the wave concomitant with the photon.

These one photon results can be extended systematically for an ensemble of N photons, using Bose-Einstein statistics and the methods of contemporary field theory, which use creation and annihilation operators. It is clear that there is an operator $\hat{B}^{(3)}$ associated with the photon in free space, and this operator is referred to henceforth as the *photomagneton* $\hat{B}^{(3)}$. It is a quantity in field theory as fundamental in nature as the operator $\hat{J}$ itself, but was identified only in 1992 [9-15]. In differential form it is defined by

$$\hat{B}^{(3)} = i \frac{B^{(0)}}{\omega} \frac{\partial}{\partial t}, \qquad (97)$$

and is linked to the one photon Hamiltonian operator $\hat{H}$ by

Chapter 3. The Origins of Wave Mechanics

$$\left|\frac{\hat{B}^{(3)}}{B^{(0)}}\right| = \left|\frac{\hat{J}}{\hbar}\right| = \frac{\hat{H}}{\hbar\omega}. \tag{98}$$

3.4 HEISENBERG UNCERTAINTY AND THE SINGLE PHOTON

Since $\hat{B}^{(3)}$, as we have just seen, is defined directly by $\hat{J}$, Heisenberg's uncertainty principle applies to it in the same way as it applies to the angular momentum of a single photon. One of the consequences of the Heisenberg uncertainty principle is that the trajectory of a particle cannot be specified. This conclusion also emerges from the classical, but relativistic, idea of a particle without mass, as described in Chap. 1. In order to define the trajectory of a particle, it is necessary to specify at each instant along its path both its position and momentum, and this is not compatible with momentum-position uncertainty [70]:

$$\delta p \delta x \geq \frac{\hbar}{2}. \tag{99}$$

The magnitude of the linear momentum of the massless photon is defined by the de Broglie relation $p = \hbar\omega/c$, so its *position cannot be specified, or "localized" in space*, either in wave mechanics or in classical special relativity. The photon without mass is not a "localized particle". Since $p = \hbar\omega/c$, however, and $\hbar$ and c are both universal constants, Lorentz invariants, the linear momentum magnitude p of the photon without mass is determined by the angular frequency ω, which is the angular frequency of a monochromatic wave. The frequency of the wave is proportional to the linear momentum of the particle, the wave frequency is specified without uncertainty, and so therefore is the light quantum, $\hbar\omega$, of energy. The energy and linear momentum of the photon without mass, propagating in free space, can be expressed in terms of psi function operators, as argued in previous sections,

$$\hat{H} = c\hat{p} = i\hbar\frac{\partial}{\partial t}, \tag{100}$$

operators which are *both* specified. Clearly therefore the particle defined by p and the monochromatic wave defined by ω *co-exist* in free space. *This is the Einstein-de Broglie interpretation of light, photon and concomitant wave co-exist in free space.* The Ψ function for one photon in free space is specified without uncertainty as the exponent $e^{-i\phi}$.

In the Copenhagen interpretation [72] the Ψ function is interpreted in terms of probability, following the original suggestion by Born referred to earlier, and expectation values such as

$$\langle \hat{A} \rangle = \int \Psi^* \hat{A} \Psi \, dV, \qquad \langle \hat{B} \rangle = \int \Psi^* \hat{B} \Psi \, dV, \qquad (101)$$

become mean values with statistical deviations

$$\Delta \hat{A} = \hat{A} - \langle \hat{A} \rangle, \qquad \Delta \hat{B} = \hat{B} - \langle \hat{B} \rangle. \qquad (102)$$

If the operators $\hat{A}$ and $\hat{B}$ form the commutator

$$[\hat{A}, \hat{B}] = i\hat{C}, \qquad (103)$$

the Heisenberg uncertainty principle is properly defined by [70]

$$\delta \hat{A} \delta \hat{B} \geq \frac{1}{2} |\langle \hat{C} \rangle|, \qquad (104)$$

where

$$\delta \hat{A} = (\langle \hat{A}^2 \rangle - \langle \hat{A} \rangle^2)^{\frac{1}{2}} = ((\Delta \hat{A})^2)^{\frac{1}{2}}, \qquad (105)$$

and similarly for $\delta \hat{B}$. Thus $\hat{A}$ and $\hat{B}$ commute (meaning $[\hat{A}, \hat{B}] = 0$) if both are specified without uncertainty. Therefore since the energy and linear momentum of the photon without mass are specified, both in classical special relativity and in the Schrödinger equation for one photon in free space, then the position of the photon in free space is unspecified completely. If the frequency of the monochromatic wave is ω its linear momentum is $p = \hbar\omega/c$, but its position is unspecified, it can exist along its propagation direction as oscillations covering the range $Z = -\infty$ to $Z = +\infty$. The monochromatic wave can be found anywhere in an infinite range, i.e., its position is completely unspecified. The probability of finding the wave at any point Z is zero, and so, in the Copenhagen interpretation, the wave is said not to co-exist with the photon without mass.

Recent double slit experiments, however [72], favor the Einstein-de Broglie interpretation, because the data show the

Chapter 3. The Origins of Wave Mechanics

simultaneous presence of photons and waves of light. The de Broglie relation interpreted in this way means that the photon's linear momentum co-exists with the frequency of the concomitant, monochromatic wave. The latter becomes the electrodynamic piloting or guiding field of the photon in free space.

There is a continuing and lively debate [72] between these two famous interpretations of quantum mechanics. The purpose of this section, however, is to introduce the newly discovered photomagneton, $\hat{B}^{(3)}$, by reference to the *angular momentum* commutator and uncertainty relations. These are relations among psi function operators which can be derived without reference to an equation of motion [70], and it is immediately clear from the definition of the photomagneton $\hat{B}^{(3)}$ as being directly proportional to the photon's angular momentum operator $\hat{J}$ (Eq.(96)) that there must exist a set of commutator relations involving the photomagneton $\hat{B}^{(3)}$ and the corresponding wave operators $\hat{B}^{(1)}$ and $\hat{B}^{(2)}$. The photon without mass in free space is the source of wave mechanics in matter, through the de Broglie wave particle dualism, and the angular momentum of the photon can neither be understood, nor properly interpreted, without the spin operator $\hat{B}^{(3)}$. The equivalent statement in classical electrodynamics is that the angular momentum of electromagnetic radiation is longitudinal, and not transverse [4], as shown in Eq.(44) of Chap. 1.

It is not difficult to show, as in the next chapter, that even within the structure of *classical* electrodynamics, Eqs. (4) can be written in commutator form

$$[\hat{B}^{(1)}, \hat{B}^{(2)}] = -iB^{(0)}\hat{B}^{(3)*}, \qquad (106)$$

with cyclic permutations of (1), (2) and (3), the coordinates of the circular basis defined in Eq. (46). This is just a matter of re-expressing the unit vectors $e^{(1)}$, $e^{(2)}$ and $e^{(3)}$ as matrices [15] such as those of Eq. (93), matrices which define three rotation generators $\hat{J}^{(1)}$, $\hat{J}^{(2)}$, and $\hat{J}^{(3)}$. The magnetic fields and rotation generators so defined are related by

$$\hat{B}^{(1)} = -B^{(0)}\hat{J}^{(1)}e^{i\phi}, \qquad \hat{B}^{(2)} = -B^{(0)}\hat{J}^{(2)}e^{-i\phi},$$

$$\hat{B}^{(3)} = iB^{(0)}\hat{J}^{(3)}, \qquad (107)$$

so that the commutative field algebra (106) is part of the

Lie algebra of the Lorentz group. This suggests that the magnetic (and electric) fields associated with the piloting wave of the photon are properties of space-time itself. It turns out [15] that magnetic fields are proportional to rotation generators and electric fields to boost generators. For our present purposes, we note that the rotation generators in space form the same commutator algebra as displayed in Eq. (106),

$$[\hat{J}^{(1)}, \hat{J}^{(2)}] = -\hat{J}^{(3)*} = \hat{J}^{(3)}, \qquad (108)$$

which becomes the more familiar

$$[\hat{J}_X, \hat{J}_Y] = i\hat{J}_Z \qquad (109)$$

in the Cartesian basis (X, Y, Z), *and which is identical, within a factor ℏ, with the commutator algebra of angular momentum operators in the wave mechanics of one massless photon*. The properties of $\hat{B}^{(1)}$, $\hat{B}^{(2)}$ and $\hat{B}^{(3)}$ can therefore be understood in terms of the well known [70] properties of angular momentum operators in wave mechanics.

In wave mechanics, the angular momentum commutator relations in a Cartesian basis are well known [70] to be

$$[\hat{J}_X, \hat{J}_Y] = i\hbar\hat{J}_Z, \qquad (110)$$

an equation which is the basis of the entire theory of angular momentum in quantum mechanics. Equation (109), for classical rotation generators, and Eq. (110) are identical within a (scalar) factor ℏ, and so rotation generator operators have the same commutator algebra as angular momentum operators in wave mechanics. Following Atkins [70], whenever operators are considered with the same three commutation relations as in Eq. (110) (and cyclic permutations), the observable described by these three operators in Euclidean space *is* an angular momentum. In this sense therefore, $\hat{B}^{(1)}$, $\hat{B}^{(2)}$ and $\hat{B}^{(3)}$ are component operators of angular momentum — that of one photon in free space. The complete, and well developed [70] theory of angular momentum in wave mechanics can therefore be applied intact to $\hat{B}^{(1)}$, $\hat{B}^{(2)}$ and $\hat{B}^{(3)}$.

In the wave mechanics of one photon in free space, therefore, the three operators $\hat{B}^{(1)}$, $\hat{B}^{(2)}$ and $\hat{B}^{(3)}$ cannot be specified simultaneously if the Heisenberg uncertainty principle is applied to them. If $\hat{B}^{(3)}$ is specified, for example, and if its expectation value is non-zero, then $\hat{B}^{(1)}$ and $\hat{B}^{(2)}$, being angular momenta in wave mechanics, cannot be specified. The classical equivalent of this statement is that the only non-zero component of angular momentum in a light beam equivalent to one photon is *longitudinal*, as in Eq. (44) of Chap. 1. The average values of the transverse angular momenta of such a beam are both zero, as discussed in Jackson [4], Chap. 6. In terms of the wave mechanics of one photon

$$\hat{B}^{(3)}|\Psi\rangle = \pm B^{(0)}|\Psi\rangle, \qquad (111)$$

where $|\Psi\rangle = e^{-i\phi}$ and $\hat{B}^{(3)} = B^{(0)}(i/\omega)(\partial/\partial t)$ specifies the angular momentum of the photon in terms of the spin field operator $\hat{B}^{(3)}$. Equation. (111) is the equation defining the angular momentum of one photon about its longitudinal axis (or coordinate) (3) in the circular basis,

$$\hat{J}^{(3)}|\Psi\rangle = M_J\hbar|\Psi\rangle, \qquad (112)$$

so that $|\Psi\rangle = e^{-i\phi}$ is identified as an *angular momentum* wave function, $|\Psi\rangle = |J, M_J\rangle$. The equation defining the field operator $\hat{B}^{(3)}$ can therefore be defined as the Schrödinger equation

$$\hat{B}^{(3)}|J, M_J\rangle = \pm B^{(0)}|J, M_J\rangle, \qquad (113)$$

where $|J, M_J\rangle = e^{-i\phi}$ is an angular momentum wavefunction of one photon in free space, and where $\hat{B}^{(3)}$ is a species of angular momentum operator whose eigenvalues are $\pm B^{(0)}$. The classical equivalent of Eq. (113) is

$$\mathbf{B}^{(3)} = \pm B^{(0)}\mathbf{k}. \qquad (114)$$

Since Eq. (113) applies to a beam of light made up of one photon, it also applies to a beam of N photons, and all the well known results of angular momentum theory in wave mechanics [70] are at our disposal.

The quantized magnetic field operators for one photon

are therefore

$$\hat{B}^{(1)} = -B^{(0)} \frac{\hat{J}^{(1)}}{\hbar} e^{i\phi}, \quad \hat{B}^{(2)} = -B^{(0)} \frac{\hat{J}^{(2)}}{\hbar} e^{-i\phi},$$

$$\hat{B}^{(3)} = iB^{(0)} \frac{\hat{J}^{(3)}}{\hbar},$$
(115)

where the $\hat{B}^{(i)}$ are now to be understood as field operators in wave mechanics. The longitudinal operator $\hat{B}^{(3)}$ is phase free and is the quantum of elementary, longitudinal, magnetic flux density in free space - the *photomagneton*. The latter is the pilot wave of photon spin angular momentum in the Einstein-de Broglie interpretation of wave mechanics. In the Copenhagen interpretation, the Heisenberg uncertainty principle applies, and if $\hat{B}^{(3)}$ is specified [70], $\hat{B}^{(1)}$ and $\hat{B}^{(2)}$ cannot be, a deduction which is consistent with the fact that the (3) component of the angular momentum of one photon is specified as the *longitudinal* eigenvalues $-\hbar$ and $\hbar$. In classical special relativity, the transverse components of the photon travelling at the speed of light are indeterminate, but the longitudinal component is determinate and relativistically invariant. Therefore the expectation value of $\hat{B}^{(3)}$ in free space is relativistically invariant also, and the specification of the operator $\hat{B}^{(3)}$ as $B^{(0)}$ multiplied by $\hat{J}$ is rigorously consistent with relativistic quantum field theory. In wave mechanics for many photons, $\hat{B}^{(3)}$ is a constant of motion [70], while $\hat{B}^{(1)}$ and $\hat{B}^{(2)}$ are governed by quantum statistics and are subject to purely quantum mechanical effects such as light squeezing [15]. The field $\hat{B}^{(3)}$, being defined by $\hat{J}$, is not subject to light squeezing, because the photon spin operator $\hat{J}$ itself is unaffected. In wave mechanics the rate of change of an expectation value of an operator is related to the commutator of $\hat{H}$ and that operator. Since $\hat{B}^{(3)}$ and $\hat{H}$ commute,

$$\frac{d}{dt}\langle\hat{B}^{(3)}\rangle = \frac{i}{\hbar}\langle[\hat{H}, \hat{B}^{(3)}]\rangle.$$
(116)

This result is consistent with the fact that $\hat{B}^{(3)}$ (being directly proportional to frequency free angular momentum), has no Planck energy (which must be proportional to frequency), and does not augment the classical electromagnetic energy density [9-15]. The expectation value, $\mathbf{B}^{(3)}$, of $\hat{B}^{(3)}$

52   Chapter 3. The Origins of Wave Mechanics

is independent of time, and its eigenvalues are specified in terms of the constant $\hbar$ and $-\hbar$. Similarly, the Stokes operator $\hat{S}_3$ is a constant of motion, so $\hat{B}^{(3)}$ is proportional to $\hat{S}_3$ [9-15]. Therefore the photomagneton $\hat{B}^{(3)}$ conserves angular momentum in free space, and this is a consequence of the isotropy of the Hamiltonian [70] in free space, and therefore a consequence of three dimensional symmetry. This is simply a way of saying that the magnitude of the spin angular momentum of the massless photon is $\pm\hbar$; and that the photomagneton $\hat{B}^{(3)}$ is a direct consequence of photon spin. The classical $\mathbf{B}^{(3)}$ is therefore a direct consequence (Eq. (4)) of the fact that there exists left and right circular (or elliptical) polarization in a light beam. The field $\hat{B}^{(3)}$ is therefore an operator generated directly from the spin of the photon in free space, and is an expectation value of Schrödinger's equation (111) for one photon. It is therefore phase free and frequency independent. Any attempt to understand the meaning of $\hat{B}^{(3)}$ must therefore be based on the meaning of the spin angular momentum operator $\hat{J}$ of one photon. Similarly, the interaction of $\hat{B}^{(3)}$ with matter is understood in the same way as that of $\hat{J}$ with matter. The total angular momentum of field and matter before and after the interaction must be the same, for example, and to understand this in wave mechanics requires the theory of angular momentum coupling because $\hat{B}^{(3)}$ is essentially an angular momentum operator of wave mechanics. This is the reason for referring to $\hat{B}^{(3)}$ as a spin field, it is obviously a property of light, it is not a conventional static magnetic field such as that generated by a solenoid, but has *all* the known properties of a magnetic flux density. If such a quantity does not act experimentally as a magnetic field, there is a fundamental inconsistency in electrodynamics.

The source of $\hat{B}^{(3)}$ is the same as that of $\hat{B}^{(1)}$ and $\hat{B}^{(2)}$ — conventionally [4] a current at infinity. Similarly, the source of photon spin is the same as that of photon energy and linear momentum. The classical relation (4) means that if any one of the three fields $\mathbf{B}^{(1)}$, $\mathbf{B}^{(2)}$, and $\mathbf{B}^{(3)}$ is zero, then so are the other two. In the quantum field theory Eq. (96) is written in the accepted notation as

$$\hat{B}^{(3)}(0; -\omega, \omega) = B^{(0)} \frac{\hat{J}}{\hbar}(0; -\omega\,\omega), \qquad (117)$$

which shows that $\hat{J}$ itself is generated from a phase free cross product of negative and positive frequency waves, or quantum mechanical wave functions. Therefore $\hat{J}$ is a stationary state of wave mechanics in the same sense as the energy of a light beam. Any interaction of $\hat{B}^{(3)}$ with matter must therefore reflect the fact that it is defined as

$$\hat{B}^{(3)} := \hat{B}^{(3)}(0; -\omega, \omega). \tag{118}$$

Similarly

$$\hat{J} := \hat{J}(0; -\omega, \omega), \quad \hat{S}_3 := \hat{S}_3(0; -\omega, \omega), \tag{119}$$

where $\hat{S}_3$ is the third Stokes operator. The description of $\hat{B}^{(3)}$ as "static" obviously refers to the fact that it has no net (i.e., explicit) functional dependence on phase because it is a stationary state in wave mechanics. Similar descriptions apply to $\hat{J}$, the angular momentum of one photon, and to the third Stokes parameter, S_3, the expectation value of the third Stokes operator $\hat{S}_3$. For a given beam intensity, the angular momentum magnitude is $\hbar$, the universal Dirac constant.

In the Copenhagen interpretation, the Heisenberg uncertainty principle applied to angular momentum results in

$$\delta\hat{B}^{(1)} \delta\hat{B}^{(2)} \geq \frac{1}{2} |B^{(0)} \hat{B}^{(3)}| \tag{120}$$

where $\delta\hat{B}^{(1)}$ and $\delta\hat{B}^{(2)}$ are root mean square deviations. As usual the right hand side is a rigorous lower bound [70] on the product $\delta\hat{B}^{(1)} \delta\hat{B}^{(2)}$, a lower bound which is therefore defined by the photomagneton $\hat{B}^{(3)}$. If $\hat{B}^{(3)}$ were zero, $\hat{B}^{(1)}$ and $\hat{B}^{(2)}$ would commute, implying $\delta\hat{B}^{(1)} = \hat{0}$ and $\delta\hat{B}^{(2)} = \hat{0}$ simultaneously. For a beam of many photons, the experimental observation [9–15] of light squeezing shows this to be inconsistent with data, therefore $\hat{B}^{(3)} \neq \hat{0}$. Such a deduction follows from the fact that $\hat{B}^{(1)}$ and $\hat{B}^{(2)}$ can be described in terms of creation and annihilation operators, which do not commute. In this sense light squeezing indicates the existence of the photomagneton $\hat{B}^{(3)}$. For one photon, the commutator $[\hat{B}^{(1)}, \hat{B}^{(2)}]$ is also non-zero, showing that these fields cannot be specified simultaneously with $\hat{B}^{(3)}$. In the same way the transverse angular momenta of one photon cannot be specified simultaneously with the longitudinal angular momentum. The only non-zero expectation value of angular momentum in a light beam is the longitudinal component, and in the wave mechanics of one photon this is the only stationary state of angular momentum. Similarly, only $\hat{B}^{(3)}$ can form a stationary state of the light beam, and time-average to a finite, non-zero, value.

CHAPTER 4. INTER-RELATION OF FIELD EQUATIONS

The emergence of the photomagneton $\hat{B}^{(3)}$ for one photon in free space emphasizes $\hat{C}$ conservation in the Maxwell equations. Application of $\hat{C}$ to the classical field $\boldsymbol{B}^{(3)}$ produces the same phase free field but with opposite sign, i.e., $B^{(0)}$ radiated by an electron is opposite in sign from that radiated by a positron, and this is in principle detectible experimentally, providing direct evidence for the anti-photon. Equation (83) of Chap. 3 has shown that for free fields, the relativistic Schrödinger, Dirac and d'Alembert equations are essentially the same, meaning that one photon in free space can be described by a Dirac equation, normally reserved for a free electron. In this chapter we explore some of the interrelations that exist between field equations for one photon, and mention along the way, some of the problems that are resolved by the discovery of $\hat{B}^{(3)}$, a discovery which has the effect of rendering field theory more consistent in classical and quantized form.

4.1 RELATION BETWEEN THE DIRAC AND D'ALEMBERT WAVE EQUATIONS

The Dirac equation of motion [46] is still widely accepted as one of the most successful relativistic generalizations of the Schrödinger equation, and is described in textbooks throughout physics and chemistry. It is much less well known, however, that the Maxwell and d'Alembert equations can be put into the same form as the Dirac equation, the Kemmer equation [46] being an example of this. This results from gauge invariance, and provides a clear description of the way in which the concepts of matter and light waves interact. Following Barut [46], in the classical picture the concepts of field and particle are distinct and different, and do not interrelate. In quantum theory, the two concepts merge as in Chap. 1, through the de Broglie wave particle dualism. In the quantum theory one field can be the source of another, one particle can be the source of another particle, concepts which are counter-intuitive and non-existent in the classical framework. In this sense a free electron can transmute into a free photon and vice versa, *one* particle, following Barut [46] is the source of the other; one field is the source of the other. The fields are relativistically described by wave functions, ψ^α which become operators in quantum field theory [46, 54]. By consideration of the classical ψ^α functions, it can be shown that the d'Alembert equation for one free photon is equivalent to a Dirac equation; one field is the source of the other, and vice-versa. This underlines the existence of the anti-photon as discussed in Chap. 2. Upon quantization, one particle (for example an electron described by the Dirac equation) becomes the source of the other (a photon described by the d'Alembert equation) and vice versa.

These ideas are based on the fundamental concepts of charge-current conservation and the conservation law of coordinate transformation known as gauge invariance, concepts which are expressions of Noether's theorem [54]. It is then conceptually possible, following Barut [46] to analyze the interaction of a Dirac field with the electromagnetic field in a classical framework. In this development, a charged particle can, for example, be described by a scalar field $\phi(x)$ and is regarded as the source of the electromagnetic field, described by the four-potential A_μ. *The reverse is also possible, the electromagnetic field can be regarded, following Barut [46], as the source of the charged particle.* Therefore, although the electromagnetic field is not a charged field, it can act as the source of a charged particle. This accords with our discussion in Chap. 2 on the $\hat{C}$ symmetry of the photon. It follows that the anti-electron (the positron) can act as the source of the *anti-field*, described by $-A_\mu$ (i.e., by A_μ with the sign of $A^{(0)}$ reversed), and that the anti-field acts as the source of the positron. Inevitably, we are led to the conclusion that the effect of $\hat{C}$ on the electromagnetic field is to form the anti-field, which upon quantization, produces the anti-photon whose properties were sketched in Chap. 2.

A more accurate development describes the charged particle by a four-component spinor field ψ satisfying the Dirac equation. This accurately describes the free electron and positron. The form of interaction between the Dirac particle (electron) and the electromagnetic field is determined by the principle of gauge invariance [46]. This leads inevitably to the introduction of A_μ to counteract the variation of the Lagrangian of the matter field under gauge transformation (i.e., under coordinate transformation). *The electromagnetic field becomes an inevitable consequence of the existence of the matter field.* This is a basic concept

The Dirac and D'Alembert Wave Equations

of contemporary gauge theory, and it follows that the electromagnetic anti-field ($-A_\mu$) is no less a direct consequence of the existence of the field of anti-matter (the positron field). Conversely, particle and anti-particle are regarded as inevitable consequences of the existence respectively of the electromagnetic field and anti-field. Therefore the law of conservation of charge results inevitably in the generation of A_μ from the matter field, here being described by a spinor field. This result can be summarized mathematically by noting that gauge invariance requires ∂_μ to be replaced as follows,

$$\partial_\mu \to D_\mu := i\partial_\mu - eA_\mu. \tag{121}$$

This equation shows that: (1) A_μ is regarded as an operator accompanying the differential operator $i\partial_\mu$; and (2) the product eA_μ is inevitably positive to charge conjugation $\hat{C}$. Therefore, if the sign of e is changed, then so is the sign of the scalar amplitude $A^{(0)}$, as discussed in Chap. 2. The change in sign of $A^{(0)}$ generates the electromagnetic *anti-field* in the same way as the change in sign of e generates the anti-*particle*. Therefore conservation of charge for the matter field results in the appearance of the gauge invariant electromagnetic field A_μ if the total Lagrangian is to remain invariant under gauge transformation of the second kind, a conservation law associated with coordinate transformation.

Having introduced these fundamental concepts, the equivalence of the Dirac and d'Alembert equations of motion can be expressed in standard notation [46] by

$$(\gamma^\mu i\partial_\mu + m)\psi = \frac{e}{c}\gamma^\mu A_\mu \psi, \tag{122}$$

where γ^μ is a Dirac matrix [54], m the mass of the charged particle (electron) and ψ a Dirac four-spinor. The left hand side of Eq. (122) is familiar in the *Dirac equation* for a free electron,

$$(\gamma^\mu i\partial_\mu + m)\psi = 0. \tag{123}$$

It follows that the Dirac equation for the free electromagnetic field is

$$\frac{e}{c}\gamma^\mu A_\mu \psi = 0, \tag{124}$$

Chapter 4. Inter-Relation of Field Equations

which can be expressed as the d'Alembert equation

$$\Box A_\mu = e\overline{\psi}\gamma_\mu\psi = 0. \tag{125}$$

Equation (124), as discussed by Holland [78], can also be expressed as

$$\frac{e}{c}\gamma^\mu \frac{\partial F}{\partial x_\mu} = 0, \tag{126}$$

where $F = (1/2) F_{\mu\nu}\gamma^{\mu\nu}$ is a bivector. Therefore A_μ operating on the four-spinor ψ produces $\partial F/\partial x_\mu$. As discussed by Holland [78], Eq. (126) can also be expressed as a Schrödinger equation,

$$i\frac{\partial F}{\partial t} = \hat{H}F, \tag{127}$$

where $\hat{H}$ is a Hamiltonian operator.

Therefore, as indicated in Eq. (83) of Chap. 3, the d'Alembert, Dirac, and Schrödinger equations of the free electromagnetic field are the same equation, that of one massless photon in vacuo. The expectation value of the novel photomagneton, the classical $B^{(3)}$, is a stationary state of the Schrödinger equation for a free photon, as discussed in Chap. 3, and is therefore a solution of the Dirac, d'Alembert, Proca, and Maxwell equations of the free photon. *It therefore has all the known properties of a magnetic field*, and in the Einstein-de Broglie interpretation is the piloting field of photon spin angular momentum.

If the free photon can be described by the Dirac equation (124), and also by the d'Alembert equation (125), then Fermi-Dirac and Bose-Einstein statistics must be interchangeable, as indeed, contemporary theory confirms [72]. This idea follows from the basic concept that the electron (a massive, spin ½ particle, obeying Fermi-Dirac statistics) is transmuted into a photon (massless, spin one particle, obeying Bose-Einstein statistics). It also follows that if the free photon is described by a Dirac equation, which is the same equation as that for a free electron (left and right hand sides of Eq. (122)), then

$$\hbar\omega = \pm\left(c^2 p^2 + m_0^2 c^4\right)^{\frac{1}{2}}, \tag{128}$$

both for the electron and for the photon. This is de

The Dirac and D'Alembert Wave Equations

Broglie's equation for matter waves, discussed in Chap. 1. If the photon is considered to have no mass, $m_0 = 0$, but for the massive photon described by the Proca equation, $m_0 \neq 0$. For the electron, m_0 is its rest mass. Equation (128) gives the possibility of negative energy states for both electron and photon. For both particles there are two positive energy eigenstates $\hbar\omega$, and two negative energy eigenstates $-\hbar\omega$. For the massless photon these two states correspond in each case to two different senses of helicity, right and left circular polarization. The prediction of the anti-photon now follows using the same line of reasoning as in the prediction of positrons. In so doing, it is necessary to postulate a Dirac sea (a vacuum [54]) made up of anti-photons, paired off according to right and left circular polarization. The Dirac equation for the free photon is no longer, in this picture, a single particle equation, because it produces both a photon and an anti-photon. Therefore ψ is regarded as a wavefunction, and the Hamiltonian operator from the Dirac equation is

$$H = \langle \overline{\psi} | i\hbar \frac{\partial}{\partial t} | \psi \rangle, \qquad (129)$$

in close formal analogy with Eq. (85) from the Schrödinger equation. The number of photons at a particular point is therefore $|\psi|^2$. In the view of Holland [78], for example, the photon as a concept is replaced entirely by a field singularity, and in this sense there is no particle of light at all. For the free photon this picture is conceptually equivalent to identifying the Dirac, d'Alembert and Schrödinger equations.

4.2 EQUATIONS OF THE QUANTUM FIELD THEORY OF LIGHT

The photon is at its most enigmatic when attempts are made to develop a rigorous procedure for the quantization of the electromagnetic field, represented by the gauge field A_μ. Despite the fact that quantum theory was initiated by Planck's hypothesis (Chap. 1) and by the introduction of the light quantum hypothesis by Einstein in 1905, the rigorously correct contemporary quantization procedures were beset with difficulties until the recent discovery of the longitudinal photomagneton $\hat{B}^{(3)}$, whose classical expectation value is $B^{(3)}$. This is the first time that the existence of a longitudinal component of electromagnetism has been recognized in free space, and because the field $B^{(3)}$ is accompanied by its dual

$-i\mathbf{E}^{(3)}/c$, there is also a pure imaginary longitudinal component of the electric field in free space. There are therefore longitudinal as well as transverse field components which must be considered in the quantization of A_μ. In contemporary terms there are three space-like components of the creation and annihilation photon operators, not two, the longitudinal component being related directly to time-like photon creation and annihilation operators [54]. The longitudinal and time-like operators, which are usually discarded as unphysical using a method such as that of Gupta and Bleuler [54], can be incorporated naturally into the theory. This is one of the most useful consequences of the discovery [9-15] of Eqs. (4).

The problems encountered with the conventional approach prior to the discovery of Eqs. (4) can be illustrated through the fact that the electromagnetic four-tensor $F_{\mu\nu}$ contains *three* electric and magnetic space-like components. If the electromagnetic field is accepted as being transverse, however, only two out of the three components are retained, and the time-like component becomes unphysical. Quantization of the gauge field A_μ (a four-vector) in the Lorentz gauge then proceeds through the well known Gupta-Bleuler method [54], in which the longitudinal and time-like quantized field states are asserted to be unphysical, despite the fact that it leads to the *physical* result

$$[\hat{a}^{(0)}(k) - \hat{a}^{(3)}(k)]|\psi\rangle = 0, \qquad (130)$$

where $\hat{a}^{(0)}(k)$ and $\hat{a}^{(3)}(k)$ denote time-like and longitudinal annihilation operators [9-15]. Ryder, for example [54] states that "..physical states are *admixtures* of longitudinal and time-like photons, such that Eq. (130) holds". The obvious problem with the Gupta-Bleuler approach is that these longitudinal and time-like states must simply be discarded as unphysical. Prior to the discovery of Eqs. (4), this arbitrary assertion had to be made in order to accord with the idea that electromagnetic waves in free space must be transverse. Equations (4), however, show that although the *waves remain transverse, they create a non-zero, real, physical, spin field* represented by $\hat{B}^{(3)}$. This is longitudinally directed in the Z (or (3)) axis, as for beam angular momentum, given classically by Eq. (44).

The existence of three space-like and one time-like components of the creation and annihilation photon operators in free space is a direct consequence of the d'Alembert equation, and the Gupta-Bleuler condition is derived from the

Equations of the Quantum Field Theory of Light 61

Lorentz condition, so there is no indication in the basic theory that longitudinal and time-like components (two of four) must be discarded as unphysical. Furthermore, phase free longitudinal electric and magnetic fields, being independent of time and divergentless, are clearly solutions of d'Alembert's equation (and of course, of the Maxwell equations) in free space. The fundamental usefulness of Eqs. (4) is that they establish, for the first time, a link between the longitudinal, phase free, spin field $\mathbf{B}^{(3)}$ (and its dual, $-i\mathbf{E}^{(3)}/c$) and the transverse, phase dependent electro-magnetic waves. This link is established through standard geometry in a circular basis, (i.e., the circular geometry of Eqs. (4)) and in terms of the conjugate product $\mathbf{B}^{(1)} \times \mathbf{B}^{(2)}$ which is simply the vectorial component of light intensity,

$$\mathbf{I} = \frac{c}{\mu_0} \mathbf{B}^{(1)} \times \mathbf{B}^{(2)}. \tag{131}$$

We now proceed to illustrate how $\mathbf{B}^{(3)}$ removes the difficulties encountered in the Gupta-Bleuler method. In the conventional approach, $\mathbf{B}^{(3)}$ is asserted, quite arbitrarily in view of Eqs. (4), to be unphysical. The happy outcome of the discovery of Eqs. (4) is that the quantization of A_μ proceeds in a more self consistent manner, indeed becomes entirely self consistent if the method of Gupta and Bleuler is accepted, a method devised to deal with the Lorentz condition [54]. This increased self consistency in the basic theory in itself gives much confidence in the physical nature of $\mathbf{B}^{(3)}$, which indeed, is observed, albeit at second order thus far [19–26], in the inverse Faraday effect, summarized in Eq. (6) of Chap. 1.

From the classical definition (Eqs. (4))

$$|\mathbf{B}^{(3)}| = B^{(0)}, \qquad |i\mathbf{E}^{(3)}| = E^{(0)}, \tag{132}$$

for the real $\mathbf{B}^{(3)}$ and the imaginary $i\mathbf{E}^{(3)}$. It is obvious that these relations are precisely analogous with Eq. (130), given the following link (S.I. units):

$$\hat{E}^{(0)} = \xi_E \hat{a}^{(0)}, \qquad \hat{E}^{(3)} = \xi_E \hat{a}^{(3)}, \qquad \hat{B}^{(0)} = \xi_B \hat{a}^{(0)},$$

$$\hat{B}^{(3)} = \xi_B \hat{a}^{(3)}, \qquad \xi_E = \left(\frac{2\hbar\omega}{\epsilon_0 V}\right)^{\frac{1}{2}}, \qquad \xi_B = \left(\frac{2\mu_0 \hbar\omega}{V}\right)^{\frac{1}{2}}. \tag{133}$$

Chapter 4. Inter-Relation of Field Equations

Equation (130), a direct result of quantization in the Lorentz gauge, is therefore seen to be

$$(\hat{B}^{(0)} - \hat{B}^{(3)})|\psi\rangle = i(\hat{E}^{(0)} - \hat{E}^{(3)})|\psi\rangle = 0, \tag{134}$$

which imply

$$\langle\psi|\hat{B}^{(0)+}\hat{B}^{(0)}|\psi\rangle = \langle\psi|\hat{B}^{(3)+}\hat{B}^{(3)}|\psi\rangle$$
$$\langle\psi|\hat{E}^{(0)+}\hat{E}^{(0)}|\psi\rangle = \langle\psi|\hat{E}^{(3)+}\hat{E}^{(3)}|\psi\rangle. \tag{135}$$

The field $\hat{B}^{(3)}$, is described in terms of a longitudinal photon annihilation operator in Eq. (133). The operator $\hat{a}^{(3)}$ is therefore phase free, like the field. The time-like photon annihilation operator $\hat{a}^{(0)}$ is also phase free. These results contrast with the usual transverse photon creation and annihilation operators, well known to be defined (in S.I.) by the phase dependent

$$\hat{a}(t) = \hat{a}(0)e^{-i\omega t}, \quad \hat{a}^+(t) = \hat{a}^+(0)e^{i\omega t}. \tag{136}$$

The question arises as to the part played by the time-like component $\hat{a}^{(0)}$ in the four-vector A_μ and electric and magnetic fields in free space. In particular, can the electric and magnetic fields be defined themselves as four-vectors? In special relativity, electric and magnetic fields are incorporated within the four-tensor $F_{\mu\nu}$ of electromagnetism, defined as the four-curl [46] of A_μ. The four-tensor is used to represent an axial quantity in four dimensions, in much the same way as an antisymmetric rank two tensor is used to represent an axial vector in the three dimensions of Euclidean space. The correct generalization of a three dimensional axial vector such as angular momentum to four dimensions is a second order antisymmetric tensor, for example angular momentum in four dimensions can be expressed as the antisymmetric four-tensor

$$J^{\mu\nu} = -J^{\nu\mu}. \tag{137}$$

The three spatial components of $J^{\mu\nu}$ represent the total, Euclidean angular momentum vector **J**. However, there are also *time-like* components J^{ok} which do not have a simple physical

Equations of the Quantum Field Theory of Light

meaning [46] for a single particle.

Therefore, methods must be found to account for the existence of longitudinal and time-like photon creation and annihilation operators (Eq. (130)) in the correct four dimensional representation of electric and magnetic fields. There are at least two ways in which this may be accomplished, i.e., ways in which the time-like components $B^{(0)}$ and $iE^{(0)}$ can be incorporated in the relativistic theory of electromagnetism. Recall that the necessity for doing this springs from Eqs. (4), which link together the transverse and longitudinal components of B in vacuo, and Eq. (132), which relates the longitudinal component $B^{(3)}$ to a time-like component $B^{(0)}$. The longitudinal and time-like creation and annihilation operators are usually discarded as unphysical using the Gupta-Bleuler method, as we have seen, but following the discovery of Eqs. (4), this can no longer be done, because $B^{(3)}$ is a *physical* magnetic field. The problem essentially reduces, therefore, to that of defining the meaning of a time-like component of the four dimensional description of an axial vector such as B. The four-tensor of electromagnetism is given by

$$F_{\mu\nu} = \epsilon_0 \begin{bmatrix} 0 & -cB^{(3)} & cB^{(2)} & -iE^{(1)} \\ cB^{(3)} & 0 & -cB^{(1)} & -iE^{(2)} \\ -cB^{(2)} & cB^{(1)} & 0 & -iE^{(3)} \\ iE^{(1)} & iE^{(2)} & iE^{(3)} & 0 \end{bmatrix}, \quad (138)$$

in which there are real and imaginary elements. In analogy with the description of angular momentum in four dimensions, Eq. (137), the real, magnetic components $cB^{(1)}$, $cB^{(2)}$ and $cB^{(3)}$ are those of the space-like Euclidean vector cB, and the imaginary $iE^{(1)}$, $iE^{(2)}$ and $iE^{(3)}$ are non space-like components of an imaginary vector iE. Barut [54] refers to the non space-like components of $J^{\mu\nu}$ as "time components". Therefore, it is reasonable to refer to iE as the "time-component" of the four-tensor $F^{\mu\nu}$, which is a representation in four dimensions of an axial vector in three dimensions. From the ordinary analytical algebra of complex numbers [79] the magnitude of the longitudinal component of the complex iE vector is $|iE^{(3)}|$. The magnitude of the space-like component is $|cB^{(3)}|$. These magnitudes are the same, because

Chapter 4. Inter-Relation of Field Equations

$$|i\boldsymbol{E}^{(3)}| = E^{(0)} = cB^{(0)} = |c\boldsymbol{B}^{(3)}|. \tag{139}$$

The results are therefore obtained that $|\boldsymbol{B}^{(3)}| = B^{(0)}$ and $|i\boldsymbol{E}^{(3)}| = E^{(0)}$, which are none other than equations (132).

This (free space) analysis shows that Eqs. (4) and Eqs. (132) are consistent with the usual definition of $F_{\mu\nu}$, and that $B^{(0)}$ can be regarded as a time-like quantity in special relativity. To throw this away as "unphysical", as in the conventional approach, is in the light of this reasoning, a quite arbitrary assertion, an assertion which runs contrary to the obviously physical nature of $\boldsymbol{B}^{(3)}$ from the cyclically symmetric Eq. (4). From Eq. (139), the real and physical $\boldsymbol{B}^{(3)}$ is related to the real and physical $B^{(0)}$. The space-like part of $F_{\mu\nu}$ is therefore $c\boldsymbol{B}$, and its time-like part is $i\boldsymbol{E}$. Each component of $c\boldsymbol{B}$ (i.e., $c\boldsymbol{B}^{(1)}$, $c\boldsymbol{B}^{(2)}$ and $c\boldsymbol{B}^{(3)}$) is space-like, each component of $i\boldsymbol{E}$ (i.e., $i\boldsymbol{E}^{(1)}$, $i\boldsymbol{E}^{(2)}$ and $i\boldsymbol{E}^{(3)}$) is time-like. The scalar magnitude of $i\boldsymbol{E}$ is therefore a time-like scalar quantity, defined by the square root of $i\boldsymbol{E}$ multiplied by its conjugate product,

$$|i\boldsymbol{E}| = (-i^2 E^2)^{\frac{1}{2}} = E. \tag{140}$$

This leads to the result that $|i\boldsymbol{E}|$ (and therefore $|c\boldsymbol{B}|$) can be regarded as time-like scalar quantities.

With this result, it is possible *in free space* to write down *formal* four-vector quantities,

$$E_\mu := (\boldsymbol{E}, iE^{(0)}), \qquad B_\mu := (\boldsymbol{B}, iB^{(0)}), \tag{141}$$

and to apply to these a formal Lorentz transformation. However, Maxwell's equations in free space are well known to be invariant to Lorentz transformation [4], and in consequence appear the same in all frames of reference. This is equivalent to saying that the frame in which the Maxwell equations are written down is not subject to Galilean relativity, so that there does not exist a frame which travels with respect to the original with a speed v. Therefore any Lorentz transformation applied to the formal, free space, four-vectors E_μ and B_μ must be carried out with $v = 0$, a process, which, of course, leaves the four-vectors unchanged. This is equivalent to saying that E_μ and B_μ can be defined only in *one* Lorentz frame, and in free space.

In summary, Eqs. (4) show beyond doubt that there

exists a phase free $B^{(3)}$ in free space, which is real and physical, and which can be related to a time-like component $B^{(0)}$ through Eq. (132). The latter is to be found in the four-tensor $F_{\mu\nu}$, and a formal definition can be constructed in free space of four-vectors E_μ and B_μ, but only in one Lorentz frame. The fact that $B^{(3)}$ magnetizes matter through the inverse Faraday effect (Eq. (6)) is experimental evidence [19–26] of its physical significance at second order. In contrast, there appears to be no such experimental evidence for the unphysical, and imaginary $iE^{(3)}$.

4.3 D'ALEMBERT AND PROCA EQUATIONS

The Proca [80] and d'Alembert equations appear at first to be identical for all practical purposes, because if m_0 denotes the (assumed) finite mass of the photon the Proca equation is

$$\Box A_\mu = -\xi^2 A_\mu, \quad \xi = \frac{m_0 c}{\hbar}, \qquad (142)$$

and is an eigenvalue equation. Numerically, ξ may be as small as $10^{-26}\ m^{-1}$, and in consequence

$$\Box A_\mu \sim 10^{-52} A_\mu \sim 0 \qquad (143)$$

for all practical purposes, the Proca and d'Alembert equations are the same, it seems. Here again, however, the argument is enigmatic, because a boson (the photon) with mass, has three, well defined, eigenvalues of spin angular momentum, $-\hbar$, 0, and $+\hbar$. This result is true however minute the numerical value of m_0 may be, and some estimates place $m_0 \sim 10^{-65}\ kgm$ or less. The existence of $m_0 \neq 0$ changes completely many aspects of electromagnetic field theory, both classical and quantum mechanical, despite the fact that m_0 is numerically so small.

(1) The Proca equation allows a three dimensional particle interpretation of the photon, in the sense that it allows minute, but physical, phase dependent longitudinal electric and magnetic fields in free space.

(2) Quantization of Eq. (142) occurs naturally in that it leads to a three dimensional particle interpre-

tation of the photon with mass.

(3) The classical relativistic field theory no longer leads to just two helicities, and the Wigner little group is no longer the physically meaningless [54] E(2).

(4) The above advantages are offset, however, by the need to rethink the basics of gauge theory, i.e., to adapt the latter for $m_0 \neq 0$ without contradicting gauge invariance of the second kind. This must be done while retaining the advantages of unified and grand unified field theory, as described for example by Huang [81], but more work is needed in this area. The problem succinctly stated is that of ensuring that the Lagrangian,

$$\mathcal{L}_m = \frac{1}{2} m_0^2 A_\mu A_\mu, \qquad (144)$$

vanishes. Conventional electromagnetic [54] and unified field theory asserts that m_0 is zero identically, but this assertion ignores experimental indications to the contrary [72], as reviewed recently by Vigier.

(5) It is clear that the condition

$$A_\mu A_\mu = 0, \qquad (145)$$

ensures that the Lagrangian vanishes for all m_0, and that Eq. (145) is also consistent with the Lorentz condition, $\partial A_\mu / \partial x_\mu = 0$, which is implied automatically by the Proca equation. Therefore, although the Proca and d'Alembert equations appear to be the same equation for all practical purposes, the former loses "gauge freedom", being valid only in the Lorentz gauge, and in no other. This rules out the use of the Coulomb gauge with the Proca equation, meaning that *solutions of the Proca equation can be longitudinal as well as transverse*. This deduction is indicated by equations (4), which link together the two types of solution, but for magnetic fields and not for the vector potential A_μ. The relevant cyclic relations for the latter are Eqs. (24), which indicate that the longitudinal part of A_μ is pure imaginary. Quantization of the electromagnetic field occurs more naturally in the Lorentz gauge [54], where it is possible to allow for the existence of an imaginary longitudinal component of A_μ as indicated by the cyclic algebra, Eqs. (24) and (25).

However, if we write the Proca equation (in Minkowski notation) as

$$\frac{\partial F_{\mu\nu}}{\partial x_\mu} = -\xi^2 A_\nu, \tag{146}$$

and multiply both sides by A_ν,

$$A_\nu \frac{\partial F_{\mu\nu}}{\partial x_\mu} = -\xi^2 A_\nu A_\nu, \tag{147}$$

it becomes clear that if $A_\mu A_\mu = 0$ the Proca equation reduces to the free space equation,

$$A_\nu \frac{\partial F_{\mu\nu}}{\partial x_\mu} = 0, \tag{148}$$

whose $m_0 = 0$ counterpart is the free space Maxwell equation [46, 54],

$$\frac{\partial F_{\mu\nu}}{\partial x_\mu} = 0. \tag{149}$$

Using the novel condition $A_\mu A_\mu = 0$ is equivalent to asserting that the four-vector A_μ is a *physically meaningful light-like vector which is manifestly covariant*. Experimental evidence for this deduction comes from the well known Aharonov-Bohm effect [82] in which an electron diffraction pattern is shifted by the vector potential associated with magnetic flux density. (It is interesting to note that if $\mathbf{B}^{(3)}$ is physically meaningful, there should also be an optical Aharonov-Bohm effect [83].)

Therefore, contemporary gauge theory can be adapted for use with finite photon mass provided that A_μ is regarded as physically meaningful and manifestly covariant, being in the light-like condition defined by Eq. (145). It can be shown [83, 84] that this condition holds to an excellent approximation, and that setting $A_\mu A_\mu$ to zero identically is equivalent to asserting that the photon radius vanishes. More accurately, use of the Dirac condition shows that $A_\mu A_\mu$ is proportional to the square of a finite photon radius, a small correction which is consistent with the fact that $A_\mu A_\mu$ deviates slightly from the exact light-like condition because of finite photon mass.

Experimental indications of finite photon mass, as discussed for example by Goldhaber and Nieto [85] and by Vigier [72], usually result in upper bounds, i.e., the photon mass is stated to be less than a given order of magnitude. However, there are so many independent experimental signs [72, 85] of the existence of light mass that there is a very high overall probability that the mass is non-zero. As soon as this is accepted, the condition $A_\mu A_\mu = 0$ becomes *the only way* to account for finite m_0 within contemporary gauge theory. It follows that A_μ must be manifestly covariant and physically meaningful if m_0 is non-zero, and so A_μ must have longitudinal components as indicated in Eqs. (24) and (25). Finally, we are led to deduce that finite photon mass actually implies the existence of longitudinal components such as $\boldsymbol{B}^{(3)}$ in free space. Experimentally the existence of $\boldsymbol{B}^{(3)}$ is consistent with the magnetization of matter by light as shown in Eq. (6). This is a straightforward chain of reasoning which links m_0 to $\boldsymbol{B}^{(3)}$. It is unsurprising in this light that $\boldsymbol{B}^{(3)}$ also emerges directly from the Proca equation (142). The proof of this result is simply obtained by writing the rigorously relativistic Proca equation in the limit,

$$\nabla^2 \boldsymbol{A} = \xi^2 \boldsymbol{A}, \tag{150}$$

a limit which gives its time independent solutions. Using the usual relation $\boldsymbol{B} = \nabla \times \boldsymbol{A}$, Eq. (150) gives the solution

$$\boldsymbol{B}^{(3)} = B^{(0)} \exp(-\xi Z) \boldsymbol{k}, \tag{151}$$

which is an exponential decay with distance, Z. As we have seen, ξ is of the order $10^{-26}\ m^{-1}$, so for all practical purposes in the laboratory, $\boldsymbol{B}^{(3)}$ from Eqs. (151) and (4) are identical. The physical justification for this procedure is that photon mass is minute, so that field solutions of the Proca and d'Alembert equations must be practically identical.

It is critically important to note, however, that $\boldsymbol{B}^{(3)}$ obtained in this way is not simply an arbitrary static Yukawa type potential [72] but is linked ineluctably with the transverse wave solutions $\boldsymbol{B}^{(1)}$ and $\boldsymbol{B}^{(2)}$ of the Proca equations. The link is established through cyclic relations of the type (4). *Longitudinal and transverse solutions of both the d'Alembert and Proca equations are linked geometrically in free space.* The next chapter illustrates the nature of this geometry.

CHAPTER 5. TRANSVERSE AND LONGITUDINAL PHOTONS AND FIELDS

Equations (4) reveal that the transverse and longitudinal components of the magnetic part of free space electromagnetism are linked together in a circular basis (1), (2) and (3), which is also a basis (Eqs. (46)) for a three dimensional geometrical representation of free space. In the first sections of this chapter the geometrical basis for magnetic and electric fields in free space is developed in terms of rotation and boost generators. This development shows that the novel spin field $B^{(3)}$ is firmly rooted in the three dimensional geometry of space and the four dimensional geometry of space-time. The field $B^{(3)}$ is therefore a fundamental magnetic field in the same sense as $B^{(1)}$ and $B^{(2)}$ are fundamental fields. Experimental evidence for this deduction is available in the inverse Faraday effect, through Eq. (6), for if $B^{(3)}$ were not a fundamental magnetic field, magnetization by light could not occur, because the quantity $M^{(3)}$ in Eq. (6) would not be a magnetization. In the same sense, force in Newton's equation would not be force if the acceleration were not an acceleration, a perfectly obvious deduction, but one which has to be emphasized in view of the novelty of $B^{(3)}$. In this respect, the electromagnetic field and the photon are at their most enigmatic, because in the conventional interpretation [4], electromagnetic plane waves are always asserted to be transverse. This assertion remains true in Eqs. (4), but the plane *waves* identified by circular coordinates (1) and (2) are accompanied inevitably by a *spin* field $B^{(3)}$, labelled by the circular coordinate (3) corresponding to the coordinate Z in the Cartesian basis. It is clear that the field $B^{(3)}$ is generated directly from an angular momentum, a pseudo-vector spinning about the (3) or Z axis.

In the light of Eq. (4), this seems a perfectly clear and obvious result, and as discussed in Chap. 2, the set of equations (4) conserve $\hat{C}$, $\hat{P}$, and $\hat{T}$. It is easily checked that the three field components $B^{(1)}$, $B^{(2)}$ and $B^{(3)}$ obey the Maxwell equations in free space and there is of course an equivalent to $B^{(3)}$ in matter. In this chapter this deduction is seen in the light of fundamental geometry to be essential-

72   Chapter 5. Transverse, Longitudinal Photons and Fields

ly geometrical in nature, and there are several ways of showing this. We start in Sec. 1 with a consideration of the fundamental unit vectors $\mathbf{i}$, $\mathbf{j}$ and $\mathbf{k}$ that define the magnetic fields in a Cartesian basis.

5.1 AXIAL UNIT VECTORS, ROTATION GENERATORS, AND MAGNETIC FIELDS

The Lie algebra of Eqs. (4) can be represented in terms of commutators of matrices, allowing a direct route to the quantization of the three Maxwellian fields $\mathbf{B}^{(1)}$, $\mathbf{B}^{(2)}$ and $\mathbf{B}^{(3)}$. The three unit vectors in the circular basis defined by Eq. (46) can be used to develop a Lie algebra of commutators in the circular basis (1), (2) and (3) rather than in the Cartesian basis X, Y, and Z. In this geometrical basis the unit vectors form the following cyclical Lie algebra:

$$\mathbf{e}^{(1)} \times \mathbf{e}^{(2)} = i\mathbf{e}^{(3)*}, \qquad \mathbf{e}^{(2)} \times \mathbf{e}^{(3)} = i\mathbf{e}^{(1)*},$$
$$\mathbf{e}^{(3)} \times \mathbf{e}^{(1)} = i\mathbf{e}^{(2)*}, \tag{152}$$

where * denotes "complex conjugation". Geometrically, if $\mathbf{e}^{(3)} = 0$, then $\mathbf{e}^{(1)} = \mathbf{e}^{(2)*} = 0$, and if $\mathbf{e}^{(3)} \neq 0$, then $\mathbf{e}^{(1)} = \mathbf{e}^{(2)*} \neq 0$. This structure is the same as that of Eqs. (4), revealing that the latter are also *geometrically* based. In other words if any of the fields $\mathbf{B}^{(1)}$, $\mathbf{B}^{(2)}$ and $\mathbf{B}^{(3)}$ is zero, so are the other two, and all electromagnetism in vacuo vanishes identically. It becomes clear that the transverse waves $\mathbf{B}^{(1)}$ and $\mathbf{B}^{(2)}$ are linked to the spin field $\mathbf{B}^{(3)}$, and *that the conventional approach to electromagnetism [4] is incomplete because it is too restrictive.*

To extend these considerations to four dimensional space-time and to quantum mechanics, it is more convenient to use commutator rather than vectorial algebra. Equations (4) can be put in commutative form by using the result from tensor analysis that an axial vector is equivalent to a rank two antisymmetric polar tensor,

$$B_k = \frac{1}{2} \epsilon_{ijk} \hat{B}_{ij}, \tag{153}$$

where ϵ_{ijk} is the Levi-Civita symbol. The rank two tensor representation of the axial vector B_k is mathematically equivalent but has the advantage of being accessible to

Axial Vectors, Rotation Generators, Magnetic Fields

commutator (matrix) algebra, allowing $\mathbf{B}^{(1)}$, $\mathbf{B}^{(2)}$ and $\mathbf{B}^{(3)}$ to be expressed as infinitesimal rotation generators and thereby as quantum mechanical angular momentum operators, as discussed in Chap. 3. These methods show that the photon has an elementary longitudinal flux quantum, the photomagneton operator $\hat{B}^{(3)}$, which is directly proportional to its intrinsic spin angular momentum.

The classical magnetic fields $\mathbf{B}^{(1)}$, $\mathbf{B}^{(2)}$ and $\mathbf{B}^{(3)}$ in vacuo are all axial vectors by definition, and it follows that their unit vector components must also be axial in nature. In matrix form, they are, in the Cartesian basis,

$$\mathbf{i} = \begin{bmatrix} 0 & 0 & 0 \\ 0 & 0 & 1 \\ 0 & -1 & 0 \end{bmatrix}, \quad \mathbf{j} = \begin{bmatrix} 0 & 0 & -1 \\ 0 & 0 & 0 \\ 1 & 0 & 0 \end{bmatrix}, \quad \mathbf{k} = \begin{bmatrix} 0 & 1 & 0 \\ -1 & 0 & 0 \\ 0 & 0 & 0 \end{bmatrix}, \tag{154}$$

and in the circular basis,

$$\hat{e}^{(1)} = \frac{1}{\sqrt{2}} \begin{bmatrix} 0 & 0 & i \\ 0 & 0 & 1 \\ -i & -1 & 0 \end{bmatrix},$$

$$\hat{e}^{(2)} = \frac{1}{\sqrt{2}} \begin{bmatrix} 0 & 0 & -i \\ 0 & 0 & 1 \\ i & -1 & 0 \end{bmatrix}, \quad \hat{e}^{(3)} = \begin{bmatrix} 0 & 1 & 0 \\ -1 & 0 & 0 \\ 0 & 0 & 0 \end{bmatrix}. \tag{155}$$

The latter form a commutator Lie algebra which is mathematically equivalent to the vectorial Lie algebra,

$$[\hat{e}^{(1)}, \hat{e}^{(2)}] = -i\hat{e}^{(3)*}, \qquad \text{and cyclic permutations.} \tag{156}$$

These are our geometrical commutators in the circular basis convenient for the electromagnetic plane wave in vacuo. Equations (4) and (156) therefore represent a closed, cyclically symmetric algebra in which all three space-like components are meaningful, and if it is arbitrarily asserted that one of these components is zero, the geometrical structure is destroyed and the algebra rendered meaningless.

The cyclical commutator basis (156) can now be used to build a matrix representation of the three space-like magnetic components of the electromagnetic plane wave in vacuo,

Chapter 5. Transverse, Longitudinal Photons and Fields

$$\hat{B}^{(1)} = iB^{(0)}\hat{e}^{(1)}e^{i\phi}, \qquad \hat{B}^{(2)} = -iB^{(0)}\hat{e}^{(2)}e^{-i\phi},$$

$$\hat{B}^{(3)} = B^{(0)}\hat{e}^{(3)},$$

(157)

from which emerges the commutative Lie algebra equivalent to the vectorial Lie algebra (4),

$$[\hat{B}^{(1)}, \hat{B}^{(2)}] = -iB^{(0)}\hat{B}^{(3)*}, \quad \text{and cyclic permutations.} \qquad (158)$$

This algebra can be expressed in terms of the infinitesimal rotation generators of O(3) in three dimensional space, rotation generators which are complex matrices [15],

$$\hat{J}^{(1)} = \frac{\hat{e}^{(1)}}{i} = \frac{1}{\sqrt{2}}\begin{bmatrix} 0 & 0 & 1 \\ 0 & 0 & -i \\ -1 & i & 0 \end{bmatrix},$$

(159)

$$\hat{J}^{(2)} = \frac{-\hat{e}^{(2)}}{i} = \frac{1}{\sqrt{2}}\begin{bmatrix} 0 & 0 & 1 \\ 0 & 0 & i \\ -1 & -i & 0 \end{bmatrix}, \quad \hat{J}^{(3)} = \frac{\hat{e}^{(3)}}{i} = \begin{bmatrix} 0 & -i & 0 \\ i & 0 & 0 \\ 0 & 0 & 0 \end{bmatrix}.$$

The magnetic field matrices and rotation generators are linked by

$$\hat{B}^{(1)} = -B^{(0)}\hat{J}^{(1)}e^{i\phi}, \qquad \hat{B}^{(2)} = -B^{(0)}\hat{J}^{(2)}e^{-i\phi},$$

$$\hat{B}^{(3)} = iB^{(0)}\hat{J}^{(3)},$$

(160)

which is a result of key importance in recognizing that the commutative algebra of the magnetic fields (158) is part of the Lie algebra of the Lorentz group of Minkowski space-time. A magnetic field is thereby shown to be a property of space-time. The real, physically meaningful, magnetic field $\hat{B}^{(3)}$ is directly proportional to the fundamental rotation generator $\hat{J}^{(3)}$, which is a fundamental property of space or space-time.

In space-time, the matrices (159) become [15] the four by four complex matrices,

Axial Vectors, Rotation Generators, Magnetic Fields

$$\hat{J}^{(1)} = \hat{J}^{(2)*} = \frac{1}{\sqrt{2}} \begin{bmatrix} 0 & 0 & 1 & 0 \\ 0 & 0 & -i & 0 \\ -1 & i & 0 & 0 \\ 0 & 0 & 0 & 0 \end{bmatrix}, \tag{161}$$

$$\hat{J}^{(2)} = \hat{J}^{(1)*} = \frac{1}{\sqrt{2}} \begin{bmatrix} 0 & 0 & 1 & 0 \\ 0 & 0 & i & 0 \\ -1 & -i & 0 & 0 \\ 0 & 0 & 0 & 0 \end{bmatrix}, \quad \hat{J}^{(3)} = -\hat{J}^{(3)*} = \begin{bmatrix} 0 & -i & 0 & 0 \\ i & 0 & 0 & 0 \\ 0 & 0 & 0 & 0 \\ 0 & 0 & 0 & 0 \end{bmatrix}.$$

It follows directly from Eq. (161) that magnetic fields in the Lorentz group are directly proportional to four by four matrices of the type (161), i.e., the magnetic components of the electromagnetic plane wave in vacuo are well defined properties of space-time. Similarly, electric components are related to boost generators of the Lorentz group, boost generators which are also four by four matrices in Minkowski space-time.

The rotation generators in space form a commutator algebra of the following type in the circular basis,

$$[\hat{J}^{(1)}, \hat{J}^{(2)}] = -\hat{J}^{(3)*}, \tag{162}$$

which becomes,

$$[\hat{J}_X, \hat{J}_Y] = i\hat{J}_Z, \tag{163}$$

in the Cartesian basis, and which is, within a factor $\hbar$, identical with the commutator algebra of angular momentum operators in quantum mechanics, as discussed in Chap. 3. This provides a simple route to quantization of the magnetic fields of the electromagnetic plane wave in vacuo, giving the result (Chap. 3)

$$\hat{B}^{(1)} = -B^{(0)} \frac{\hat{J}^{(1)}}{\hbar} e^{i\phi}, \quad \hat{B}^{(2)} = -B^{(0)} \frac{\hat{J}^{(2)}}{\hbar} e^{-i\phi},$$

$$\hat{B}^{(3)} = iB^{(0)} \frac{\hat{J}^{(3)}}{\hbar}, \tag{164}$$

where $\hat{B}^{(i)}$ are now operators in quantum mechanics. In particular, the longitudinal $\hat{B}^{(3)}$ is the photomagneton opera-

Chapter 5. Transverse, Longitudinal Photons and Fields

tor, which is a stationary state in wave mechanics as discussed in Chap. 3. This result is in precise agreement with the original derivation [9-15] using the Stokes operator $\hat{S}_3$, which can be defined in terms of creation and annihilation operators as described by Tanaś and Kielich [86].

These results can be generalized (Sec. 5.2) to electric fields using boost generators, $\hat{K}^{(i)}$, which in the Lorentz group are also 4 x 4 matrices,

$$\hat{E}^{(1)} = E^{(0)} \hat{K}^{(1)} e^{i\phi}, \quad \hat{E}^{(2)} = E^{(0)} \hat{K}^{(2)} e^{-i\phi},$$

$$i\hat{E}^{(3)} = iE^{(0)} \hat{K}^{(3)}.$$

(165)

Therefore electric fields are boost generators, magnetic fields are rotation generators. It follows that the Lie algebra of electric and magnetic fields in space-time is isomorphic with that of the infinitesimal generators of the Lorentz group of special relativity. The latter type of Lie algebra can be summarized as,

$$[\hat{J}^{(1)}, \hat{J}^{(2)}] = -\hat{J}^{(3)*}, \quad \text{and cyclic permutations,}$$

$$[\hat{K}^{(1)}, \hat{K}^{(2)}] = -i\hat{e}^{(3)*}, \quad " \quad " \quad " \quad ,$$

$$[\hat{K}^{(1)}, \hat{e}^{(2)}] = -i\hat{K}^{(3)*}, \quad " \quad " \quad " \quad ,$$

$$[\hat{K}^{(1)}, \hat{J}^{(1)}] = 0, \quad \text{etc.}$$

(166)

This isomorphism is conclusive evidence for the existence of the longitudinal fields $B^{(3)}$ and $iE^{(3)}$ in free space, because these components are isomorphic with the longitudinal rotation and boost generators which are fundamental infinitesimal generators of the Lorentz group.

It is also significant that the rotation and boost generators of the Lorentz group are isomorphic (Chap. 6) with a field algebra [87] consisting of bilinear products of creation and annihilation operators,

Axial Vectors, Rotation Generators, Magnetic Fields

$$[\hat{J}_X, \hat{J}_Y] = i\hat{J}_Z, \quad [\hat{K}_X, \hat{K}_Y] = -i\hat{J}_Z, \quad [\hat{J}_X, \hat{K}_Y] = iK_Z, \quad [\hat{J}_i, \hat{K}_i] = 0,$$

$$\hat{J}_X = -i(\hat{a}_Y^+\hat{a}_Z - \hat{a}_Z^+\hat{a}_Y), \quad \hat{K}_X = -i(\hat{a}_0^+\hat{a}_X - \hat{a}_X^+\hat{a}_0),$$

$$\hat{J}_Y = -i(\hat{a}_Z^+\hat{a}_X - \hat{a}_X^+\hat{a}_Z), \quad \hat{K}_Y = -i(\hat{a}_0^+\hat{a}_Y - \hat{a}_Y^+\hat{a}_0), \tag{167}$$

$$\hat{J}_Z = -i(\hat{a}_X^+\hat{a}_Y - \hat{a}_Y^+\hat{a}_X), \quad \hat{K}_Z = -i(\hat{a}_0^+\hat{a}_Z - \hat{a}_Z^+\hat{a}_0).$$

These bilinear products consist of longitudinal and time-like creation and annihilation operators interpreted physically in Chap. 4. The algebra (167) therefore shows that the infinitesimal rotation and boost generators of the Lorentz group of special relativity can be determined fully only if longitudinal and time-like creation and annihilation operators are used as well as the conventional transverse operators [88]. If the longitudinal and time-like components were not physically meaningful the Lie algebra (167) would collapse, in the same way as the Lie algebra represented by Eq. (4) would collapse if $\hat{B}^{(3)}$ were set to zero or otherwise taken to be unphysical.

5.2 POLAR UNIT VECTORS, BOOST GENERATORS, AND ELECTRIC FIELDS

An electric field is a polar vector in three, Euclidean, dimensions, and unlike an axial vector, cannot be put into a 3 x 3 matrix form such as embodied in Eq. (154). The cross product of two polar vectors is, however, an axial vector in Euclidean space. For example, the product

$$\boldsymbol{i} \times \boldsymbol{j} = \boldsymbol{k}, \tag{168}$$

produces the Cartesian, axial, unit vector $\boldsymbol{k}$, which in the circular basis is $\boldsymbol{e}^{(3)}$. In Minkowski space-time the axial vector $\boldsymbol{k}$, as described in Sec. 5.1, becomes a 4 x 4 matrix, related directly to the infinitesimal rotation generator $\hat{J}^{(3)}$ of the Lorentz group. It follows that a rotation generator in space-time is the result of a classical commutation of two matrices which play the role of polar vectors. From the well established Lie algebra (166) or (167) of the Lorentz group these are infinitesimal *boost* generators, 4 x 4 real matrices. The equivalent of Eq. (168) in Minkowski space-time is therefore

Chapter 5. Transverse, Longitudinal Photons and Fields

$$[\hat{K}_X, \hat{K}_Y] = -i\hat{J}_Z, \qquad (169)$$

and cyclic permutations. In the circular basis (1), (2) and (3) rather than in the Cartesian basis X, Y, and Z, this commutator algebra becomes

$$[\hat{K}^{(1)}, \hat{K}^{(2)}] = -i\hat{e}^{(3)*}, \qquad (170)$$

and cyclic permutations of (1), (2) and (3). Therefore, although polar vectors cannot be put in a matrix form in Euclidean space, they correspond to boost generators, 4 x 4 matrices, in Minkowski space-time.

Therefore, this geometrical result leads to the conclusion that electric fields in space-time are proportional to boost generators because electric fields in Euclidean space are proportional to polar unit vectors. In Euclidean space, electric field solutions of Maxwell's equations are conventionally regarded as the transverse, oscillatory, fields,

$$\boldsymbol{E}^{(1)} = \frac{E^{(0)}}{\sqrt{2}}(\boldsymbol{i} - i\boldsymbol{j})e^{i\phi}, \qquad \boldsymbol{E}^{(2)} = \frac{E^{(0)}}{\sqrt{2}}(\boldsymbol{i} + i\boldsymbol{j})e^{-i\phi}, \qquad (171)$$

which can be written directly in terms of the unit vectors of the circular basis,

$$\boldsymbol{E}^{(1)} = E^{(0)} \boldsymbol{e}^{(1)} e^{i\phi}, \qquad \boldsymbol{E}^{(2)} = E^{(0)} \boldsymbol{e}^{(2)} e^{-i\phi}. \qquad (172)$$

In Minkowski space-time, the equivalents are therefore

$$\hat{E}^{(1)} = E^{(0)} \hat{K}^{(1)} e^{i\phi}, \qquad \hat{E}^{(2)} = E^{(0)} \hat{K}^{(2)} e^{-i\phi}. \qquad (173)$$

The phase ϕ, as we have seen in Chap. 1, is a Lorentz invariant, and remains the same in space-time and Euclidean space. The boost generators appearing in Eq. (173), are written in a circular basis,

Polar Vectors, Boost Generators, Electric Fields

$$\hat{K}^{(1)} = \frac{1}{\sqrt{2}} \begin{bmatrix} 0 & 0 & 0 & 1 \\ 0 & 0 & 0 & -i \\ 0 & 0 & 0 & 0 \\ -1 & i & 0 & 0 \end{bmatrix}, \quad \hat{K}^{(2)} = \frac{1}{\sqrt{2}} \begin{bmatrix} 0 & 0 & 0 & 1 \\ 0 & 0 & 0 & i \\ 0 & 0 & 0 & 0 \\ -1 & -i & 0 & 0 \end{bmatrix}, \quad (174)$$

and correspond to the complex, polar, unit vectors $e^{(1)}$ and $e^{(2)}$ in Euclidean space.

By reference to the Lie commutator algebra (170) it is clear that the commutation of $\hat{K}^{(1)}$ and $\hat{K}^{(2)}$ is $\hat{J}^{(3)} = -\hat{J}^{(3)*}$, a *rotation* generator, directly proportional to a *magnetic* field. The equivalent result in Euclidean space is

$$\hat{e}^{(1)} \times \hat{e}^{(2)} = i\hat{e}^{(3)}. \quad (175)$$

It is not possible to form a real electric field from the cross product of $E^{(1)}$ and $E^{(2)}$ and this conforms with fundamental symmetry, as described in Chap. 2, where it was also argued that an imaginary electric field has the axial symmetry of a real magnetic field. This is an important consideration when dealing with the question of what is the electric field proportional to the third boost generator $\hat{K}^{(3)}$ of Minkowski space-time. The relevant cyclic electric field algebra in three dimensions is that of Eq. (25c). By writing out the longitudinal rotation and boost generators,

$$\hat{J}^{(3)} = \begin{bmatrix} 0 & -i & 0 & 0 \\ i & 0 & 0 & 0 \\ 0 & 0 & 0 & 0 \\ 0 & 0 & 0 & 0 \end{bmatrix}, \quad \hat{K}^{(3)} = \begin{bmatrix} 0 & 0 & 0 & 0 \\ 0 & 0 & 0 & 0 \\ 0 & 0 & 0 & 1 \\ 0 & 0 & -1 & 0 \end{bmatrix}, \quad (176)$$

it is seen that the former is pure imaginary and that the latter is pure real. It follows that either:

(1) the longitudinal $\hat{B}^{(3)}$ is pure real and the longitudinal $i\hat{E}^{(3)}$ is pure imaginary;
(2) or vice versa.

Choice (1) follows, however, from a consideration of the nature of the unit vectors $e^{(1)}$, $e^{(2)}$, and $e^{(3)}$ of the circular basis (46), in which the axial $e^{(3)}$ is pure real and equal to the real, axial, Cartesian k. From Eq. (41), multiplying this real, axial, vector by the amplitude $B^{(0)}$ (a real scalar) gives a *real* **B**$^{(3)}$ in Euclidean space, and a real $\hat{B}^{(3)}$ in space-time. This real $\hat{B}^{(3)}$ is therefore defined as

$$\hat{B}^{(3)} = iB^{(0)}\hat{J}^{(3)}, \qquad (177)$$

in terms of the imaginary rotation generator $\hat{J}^{(3)}$. It follows that the imaginary $i\hat{E}^{(3)}$ must be defined as

$$i\hat{E}^{(3)} = iE^{(0)}\hat{K}^{(3)}, \qquad (178)$$

in terms of the real boost generator $\hat{K}^{(3)}$.

5.3 LIE ALGEBRA OF ELECTRIC AND MAGNETIC FIELDS IN THE LORENTZ GROUP, ISOMORPHISM

The complete Lie algebra of the infinitesimal boost and rotation generators of the Lorentz group can be written as we have seen in either a circular basis, Eq. (166), or a Cartesian basis, Eq. (167). In matrix form, the generators are

$$\hat{K}_X = \begin{bmatrix} 0 & 0 & 0 & 1 \\ 0 & 0 & 0 & 0 \\ 0 & 0 & 0 & 0 \\ -1 & 0 & 0 & 0 \end{bmatrix}, \quad \hat{K}_Y = \begin{bmatrix} 0 & 0 & 0 & 0 \\ 0 & 0 & 0 & 1 \\ 0 & 0 & 0 & 0 \\ 0 & -1 & 0 & 0 \end{bmatrix}, \quad \hat{K}_Z = \begin{bmatrix} 0 & 0 & 0 & 0 \\ 0 & 0 & 0 & 0 \\ 0 & 0 & 0 & 1 \\ 0 & 0 & -1 & 0 \end{bmatrix},$$

$$\hat{J}_X = \begin{bmatrix} 0 & 0 & 0 & 0 \\ 0 & 0 & -i & 0 \\ 0 & i & 0 & 0 \\ 0 & 0 & 0 & 0 \end{bmatrix}, \quad \hat{J}_Y = \begin{bmatrix} 0 & 0 & i & 0 \\ 0 & 0 & 0 & 0 \\ -i & 0 & 0 & 0 \\ 0 & 0 & 0 & 0 \end{bmatrix}, \quad \hat{J}_Z = \begin{bmatrix} 0 & -i & 0 & 0 \\ i & 0 & 0 & 0 \\ 0 & 0 & 0 & 0 \\ 0 & 0 & 0 & 0 \end{bmatrix}.$$
(179)

The relation between fields and generators in space-time can be summarized as

$$\hat{B}^{(1)} = -B^{(0)}\hat{J}^{(1)}e^{i\phi} = iB^{(0)}\hat{e}^{(1)}e^{i\phi},$$

$$\hat{B}^{(2)} = -B^{(0)}\hat{J}^{(2)}e^{-i\phi} = -iB^{(0)}\hat{e}^{(2)}e^{-i\phi}, \qquad (180)$$

$$\hat{B}^{(3)} = iB^{(0)}\hat{J}^{(3)} = B^{(0)}\hat{e}^{(3)},$$

$$\hat{E}^{(1)} = E^{(0)}\hat{K}^{(1)}e^{i\phi}, \quad \hat{E}^{(2)} = E^{(0)}\hat{K}^{(2)}e^{-i\phi}, \quad i\hat{E}^{(3)} = iE^{(0)}\hat{K}^{(3)},$$

Lie Algebra of Fields in the Lorentz Group

leading to a Lie algebra,

$$[\hat{B}^{(1)}, \hat{B}^{(2)}] = iB^{(0)}\hat{B}^{(3)*}, \quad \text{and cyclic permutations,}$$

$$[\hat{E}^{(1)}, \hat{E}^{(2)}] = iE^{(0)2}\hat{e}^{(3)*}, \quad " \quad " \quad " \quad ,$$

$$[\hat{E}^{(1)}, \hat{B}^{(2)}] = iB^{(0)}(i\hat{E}^{(3)}), \quad " \quad " \quad " \quad , \tag{181}$$

$$[\hat{E}^{(1)}, \hat{B}^{(1)}] = 0, \quad \text{etc.,}$$

where we have used the notation, $i\hat{e}^{(1)*} = \hat{J}^{(1)*}$, $-i\hat{e}^{(2)*} = \hat{J}^{(2)*}$, $i\hat{e}^{(3)*} = \hat{J}^{(3)*}$, $i\hat{e}^{(2)} = \hat{J}^{(2)}$, $-i\hat{e}^{(1)} = \hat{J}^{(1)}$, $i\hat{e}^{(3)} = -\hat{J}^{(3)}$. This Lie algebra is obviously destroyed by the usual (and arbitrary) assertion that $\hat{B}^{(3)}$ is zero. Although $i\hat{E}^{(3)}$ is imaginary, it too is rigorously non-zero as we have seen.

Therefore the Lie algebra of the magnetic and electric components of plane waves and spin fields in free space is isomorphic with that of the infinitesimal boost and rotation generators of the Lorentz group of space-time. The magnetic and electric components are also interconvertible through the duality transformation of special relativity. Experimental evidence suggests that the spin field $\hat{B}^{(3)}$ is real and physical, and the duality transformation therefore implies that the longitudinal electric component is imaginary and unphysical. The isomorphism of the Lie algebra (181) and (166) indicates that there exists a one to one correspondence between all the elements of the groups. This result implies that the theory of electromagnetism in free space is relativistically rigorous if and only if the longitudinal fields $\hat{B}^{(3)}$ and $i\hat{E}^{(3)}$ are accounted for through the appropriate algebra. If $\hat{B}^{(3)}$ and $i\hat{E}^{(3)}$ are set to zero, as in the conventional approach [4], then *the isomorphism is lost* meaning that electromagnetism in free space becomes incompatible with special relativity. Setting $\hat{B}^{(3)}$ to zero, for example, has the logical consequence that the rotation generator $\hat{J}^{(3)}$ is zero, an incorrect result. Similarly, setting $i\hat{E}^{(3)}$ to zero reduces the boost generator $\hat{K}^{(3)}$ to zero incorrectly. (Setting **B**$^{(3)}$ to zero has the obvious result of reducing **M**$^{(3)}$ to zero in Eq. (6), and this is incompatible with experimental data. Similarly, experimentally observed [9–15] light shifts in atomic spectra due to **B**$^{(3)}$ would disappear.)

The conventional theory of electromagnetism (e.g. Ref. 4) has missed the existence of the field **B**$^{(3)}$ in vacuo because the relation (4) and developments thereof was not identified until recently [9–15]. It is important to realize that these relations link together the spin and wave fields in such a way that the existence of one implies the other: if one is set to zero arbitrarily the other vanishes and we lose all electromagnetism. It is also important to realize that the conjugate product $B^{(1)} \times B^{(2)}$, although a pure imaginary quantity, is made up of a pure *real* $B^{(3)}$ multiplied by $iB^{(0)}$. The conjugate product can, for example, form a *real* interaction Hamiltonian by multiplication with another imaginary quantity (a susceptibility), or a *real* magnetic dipole moment by multiplication by an imaginary hyperpolarizability as shown by Woźniak et al. [15]. Although imaginary in nature, the conjugate product is therefore a *physical* quantity, and is the antisymmetric part of light intensity (Eq. (131)). Obviously, the conjugate product *per se* is not an electric or magnetic field, being the product of two fields, and the rule that an imaginary field is unphysical, a real field is physical, cannot be applied to the conjugate product. The latter owes its existence to the fact that it is the cross product of a complex field with its conjugate, the complex nature of the field being derived from that fact that it is the solution of Maxwell's equations in vacuo. If the field were not complex, then the cross product of that field with its own conjugate would always vanish identically, and there would be no antisymmetric part of light intensity, contradicting experimental data as in antisymmetric light scattering.

In this chapter it has been shown that $B^{(3)}$ and its dual $-iE^{(3)}/c$ are essential to retain isomorphism with the Lorentz group. In other words, geometry is contradicted if $B^{(3)}$ and $-iE^{(3)}/c$ vanish because the infinitesimal generators $\hat{J}^{(3)}$ and $\hat{K}^{(3)}$ would vanish. This point is emphasized in the next section by the use of vectorial spherical harmonics.

5.4 THE EIGENVALUES OF THE MASSLESS AND MASSIVE PHOTONS: VECTOR SPHERICAL HARMONICS AND IRREDUCIBLE REPRESENTATIONS OF LONGITUDINAL FIELDS

In units of $\hbar$, the reduced Planck constant, the eigenvalues of the massless photon are -1 and +1, and those of the massive photon are -1, 0, and +1. In Euclidean space the latter are obtained from eigenequations of the type

Eigenvalues of the Massless and Massive Photons

$$\hat{J}^{(3)} e^{(1)} = +1 e^{(1)}, \qquad \hat{J}^{(3)} e^{(2)} = -1 e^{(2)}, \tag{182}$$

$$\hat{J}^{(3)} e^{(3)} = 0 e^{(3)},$$

where $\hat{J}^{(3)}$ is the rotation operator,

$$\hat{J}^{(3)} = i e^{(3)} \times = \begin{bmatrix} 0 & -i & 0 \\ i & 0 & 0 \\ 0 & 0 & 0 \end{bmatrix}. \tag{183}$$

There is no paradox in the use of $e^{(3)}$ as an operator as well as a unit vector, in the same sense that there is no paradox in the use of the scalar spherical harmonics as operators. This is discussed for example by Silver [89]. The rotation operators in Euclidean space are first rank $\hat{T}$ operators, which are irreducible tensor operators and under rotations transform into linear combinations of each other. The $\hat{T}$ operators are directly proportional to the scalar spherical harmonic operators. The rotation operators, $\hat{J}$, of the full rotation group are related to the $\hat{T}$ operators as follows,

$$\hat{T}^1_{-1} = i\hat{J}^{(1)}, \qquad \hat{T}^1_1 = i\hat{J}^{(2)}, \qquad \hat{T}^1_0 = i\hat{J}^{(3)}, \tag{184}$$

and to the scalar spherical harmonic operators by

$$\hat{Y}^1_{-1} = \frac{i}{r}\left(\frac{3}{4\pi}\right)^{\frac{1}{2}} \hat{J}^{(1)}, \qquad \hat{Y}^1_1 = \frac{i}{r}\left(\frac{3}{4\pi}\right)^{\frac{1}{2}} \hat{J}^{(2)},$$

$$\hat{Y}^1_0 = \frac{i}{r}\left(\frac{3}{4\pi}\right)^{\frac{1}{2}} \hat{J}^{(3)}. \tag{185}$$

This implies in turn that the fields $\hat{B}^{(1)}$, $\hat{B}^{(2)}$ and $\hat{B}^{(3)}$ are also operators of the full rotation group, and are therefore *irreducible representations of the full rotation group*. Specifically,

$$\hat{B}^{(1)} = B^{(0)} r \left(\frac{4\pi}{3}\right)^{\frac{1}{2}} \hat{Y}^1_{-1} e^{i\phi}, \tag{186a}$$

$$\hat{B}^{(2)} = B^{(0)} r \left(\frac{4\pi}{3}\right)^{\frac{1}{2}} \hat{Y}^1_1 e^{-i\phi}, \tag{186b}$$

84   Chapter 5. Transverse, Longitudinal Photons and Fields

$$\hat{B}^{(3)} = B^{(0)} r \left(\frac{2\pi}{3}\right)^{\frac{1}{2}} \hat{Y}_0^1, \qquad (186c)$$

which shows that $\hat{B}^{(3)}$ =? $\hat{0}$ *violates the fundamentals of group theory*. Essentially, Eqs. (186) represent $\hat{B}^{(1)}$, $\hat{B}^{(2)}$ and $\hat{B}^{(3)}$ in spherical polar coordinates (r, θ, ϕ) where ϕ in this context should not be confused with the phase ϕ of the plane wave. Therefore $\hat{B}^{(1)}$, $\hat{B}^{(2)}$ and $\hat{B}^{(3)}$ in operator form are all non-zero components of the same rank one scalar spherical harmonic Y_M^1, $M = -1, 0, +1$. Furthermore, since the operators $\hat{J}^{(1)}$, $\hat{J}^{(2)}$ and $\hat{J}^{(3)}$ are components in a circular basis of the spin, or intrinsic, angular momentum of the vector field representing the electromagnetic field, the fields $\hat{B}^{(1)}$, $\hat{B}^{(2)}$ and $\hat{B}^{(3)}$ themselves are components of spin angular momentum, as we have seen in Chaps. 3 and 4. It is also clear that $\hat{J}^{(1)}$ is a lowering (annihilation) operator,

$$\hat{J}^{(1)} e^{(2)} = +1 e^{(3)}, \quad \hat{J}^{(1)} e^{(3)} = -1 e^{(1)}, \quad \hat{J}^{(1)} e^{(1)} = 0 e^{(2)}, \quad (187)$$

and that $\hat{J}^{(2)}$ is a raising (creation) operator,

$$\hat{J}^{(2)} e^{(2)} = 0 e^{(1)}, \quad \hat{J}^{(2)} e^{(3)} = -1 e^{(2)}, \quad \hat{J}^{(2)} e^{(1)} = +1 e^{(3)}, \quad (188)$$

The total angular momentum J^2 is also an eigenoperator, for example,

$$J^2 e^{(3)} = l(l+1) e^{(3)}, \quad l = 1. \qquad (189)$$

The rotation operators therefore operate on the unit vectors $e^{(1)}$, $e^{(2)}$ and $e^{(3)}$. The operator $\hat{J}^{(3)}$ is therefore also an intrinsic spin and can be identified in the quantum theory as an intrinsic spin of the massive photon, with eigenvalues $-\hbar$, 0, and $\hbar$, or of the massless photon, with eigenvalues $-\hbar$ and $\hbar$.

For a classical vector field, its intrinsic (spin) angular momentum is identifiable with its transformation properties under rotations, and within a factor $\hbar$ the rotation operators $\hat{J}$ are spin angular momentum operators of the spin one boson. Recognition of a non-zero $\hat{B}^{(3)}$ is therefore compatible with the eigenvalues of both the massive and massless bosons. The vector spherical harmonics [89] are specific vector fields which are eigenvalues of j^2 and of $\hat{j}_z$, where $\hat{j}$ is the operator for vector fields of infinitesimal rotations about axis (3). They have definite total angular momentum and occur in sets of dimension $(2j+1)$ which span in standard form the D representations of the full rotation group, and are therefore irreducible tensors of rank j. Defining the total angular momentum as the sum of the "orbital" angular momentum $\hat{l}$ and intrinsic (spin) angular momentum $\hat{J}$, we have,

$$\hat{j} = \hat{l} + \hat{J}, \qquad (190)$$

and the vector spherical harmonics are compound irreducible tensor operators [89],

$$\hat{Y}^L_{M11} := [\hat{Y}^1 \otimes \hat{e}]^L_M. \qquad (191)$$

They are formed from scalar spherical harmonics $\hat{Y}^1_m$, which form a complete set for scalar functions, and the $\hat{e}^{(i)}$ operators, which form a complete set for any vector in three dimensional Euclidean space. Therefore the vector spherical harmonics form a complete set for the expansion of any arbitrary classical vector field,

$$\mathbf{A} = A_X \mathbf{i} + A_Y \mathbf{j} + A_Z \mathbf{k}, \qquad (192)$$

in a Cartesian basis. For this vector the $\hat{l}_z$ operates on the A_X, A_Y and A_Z. The operator $\hat{J}_z$ on $\mathbf{i}$, $\mathbf{j}$ and $\mathbf{k}$. Therefore $\hat{l}_z$ operates on the spatial part of the field, and $\hat{J}_z (= \hat{J}^{(3)})$ on the vector part.

Therefore, the operator for infinitesimal rotations about the Z axis contains two "angular momentum" operators, $\hat{l}$ and $\hat{J}$, analogous to orbital and spin angular momentum in the quantum theory of atoms and molecules. The infinitesimal rotation is therefore formally a coupling of a set of spatial fields transforming according to $D^{(1)}$ with a set of three vector fields, ($\mathbf{e}^{(1)}$, $\mathbf{e}^{(2)}$, $\mathbf{e}^{(3)}$) transforming according to $D^{(1)}$. Equation (191) is an expression of this coupling, or combining, of entities in two different spaces to give a total angular momentum. It follows from these considerations that the vector spherical harmonics are defined by

Chapter 5. Transverse, Longitudinal Photons and Fields

$$Y^L_{M11} = \sum_{mn} \langle 11mn | 11LM \rangle Y^1_m e_M, \qquad (193)$$

where $\langle 11mn|11LM \rangle$ are Clebsch-Gordan, or coupling, coefficients [89]. For photons, regarded as bosons of unit spin, it is possible to multiply Eq. (193) by $\langle 110M|11LM \rangle$ and to sum over L [89]. Using the orthogonality condition

$$\sum_j \langle j_1 m'_1 j_2 m - m'_1 | j_1 j_2 jm \rangle \langle j_1 j_2 jm | j_1 m_1 j_2 m - m_1 \rangle = \delta_{m_1 m'_1} \qquad (194)$$

it is found that

$$Y^1_0(\theta, \phi) e_M = \sum_{L=|l-1|}^{l+1} \langle 110M | 11LM \rangle Y^L_{M11}, \qquad (195)$$

which is an expression for the unit vectors e_M in terms of sums over vector spherical harmonics, i.e., of irreducible compound tensors, representations of the full rotation group of Euclidean space.

It is usually asserted in the conventional approach [4] that in the theory of free space electromagnetism, the transverse components of e_M are physical, but the longitudinal component corresponding to $M = 0$ is unphysical. This deduction corresponds to only two states of polarization, left and right circular, in the classical theory of light. However, we see from our previous considerations that this assertion amounts to $e_0 := e^{(3)} =? 0,$, meaning the incorrect disappearance of some vector spherical harmonics which are non-zero from fundamental group theory because some *irreducible* representations are incorrectly set to zero. This point can be emphasized [89] by expanding $B^{(3)}$ in terms of Wigner 3-j symbols, which gives results such as

$$B^{(3)} = B^{(0)} e^{(3)} = 2B^{(0)} \frac{Y^1_{001}}{Y^1_0} = \frac{B^{(0)}}{\sqrt{3}} \frac{(\sqrt{2} Y^2_{001} - Y^0_{011})}{Y^1_0}, \qquad (196)$$

which shows that $B^{(3)}$ is non-zero and proportional to the vector spherical harmonic Y^1_{001}, which is of course also non-zero. There is no way of asserting that $B^{(3)}$ is zero without destroying the fundamentals of group theory.

Since all three of $e^{(1)}$, $e^{(2)}$ and $e^{(3)}$ can be expressed in terms of vector spherical harmonics, they are linked linearly, as well as non-linearly, in Euclidean space. Thus, in addition to the non-linear equations (4) we have the linear set,

$$\boldsymbol{B}^{(3)} = B^{(0)} \boldsymbol{e}^{(3)} = \frac{\sqrt{2}}{2} a B^{(0)} (\boldsymbol{e}^{(1)} + \boldsymbol{e}^{(2)}) + B^{(0)} \boldsymbol{b}$$

$$= -\frac{\sqrt{2}}{2} c B^{(0)} (\boldsymbol{e}^{(1)} - \boldsymbol{e}^{(2)}) + B^{(0)} \boldsymbol{d},$$

(197)

where the coefficients are defined by the following combinations of scalar and vector spherical harmonics:

$$a = \frac{2}{\sqrt{2}} \left(\frac{Y_0^1}{Y_1^1 - Y_{-1}^1} \right), \quad c = -\frac{2}{\sqrt{2}} \left(\frac{Y_0^1}{Y_1^1 + Y_{-1}^1} \right),$$

$$b = \sqrt{2} \left(\frac{Y_{111}^1 + Y_{-111}^1}{Y_1^1 - Y_{-1}^1} \right), \quad d = \sqrt{2} \left(\frac{Y_{111}^1 - Y_{-111}^1}{Y_1^1 + Y_{-1}^1} \right),$$

This result shows that $\boldsymbol{B}^{(3)}$ is not zero because $\boldsymbol{B}^{(1)}$ and $\boldsymbol{B}^{(2)}$ are not zero.

The standard theory, as described by Silver (Chap. 29 of Ref. 89) concerns itself with the multipole expansion of a plane wave in terms of the vector spherical harmonics. In this approach, there are considered to be only two, physically significant, values of M in Eq (195), corresponding to $M = +1$ and -1, which translates into our notation as follows,

$$\boldsymbol{e}_1 := -\boldsymbol{e}^{(2)}, \quad \boldsymbol{e}_{-1} := \boldsymbol{e}^{(1)}, \quad \boldsymbol{e}_0 := \boldsymbol{e}^{(3)}, \quad (198)$$

In our new approach, which considers the experimentally proven existence of $\boldsymbol{B}^{(3)}$, the case $M = 0$ is also considered to be physically meaningful. In consequence, there is an additional pure real magnetic 2^L-pole component of the electromagnetic plane wave in vacuo corresponding to $\boldsymbol{B}^{(3)}$. The longitudinal 2^L-pole electric component is pure imaginary from fundamental considerations, as we have seen from previous chapters, but is also non-zero. The vector spherical harmonics Y_{ML1}^L, with $l = L$, are no longer transverse fields and the vector $\boldsymbol{e}^{(3)}$, which is longitudinal, can also be expressed in terms of the $L = 1$, $M = 0$ vector spherical harmonics as in Eq. (196). This result augments arguments such as that leading to Silver's [89] Eq. (28.15), where are displayed other, transverse, combinations of vector spherical harmonics. Equation (196) now shows that the longitudinal $\mathbf{e}^{(3)}$ (and therefore the longitudinal $\mathbf{B}^{(3)}$) can be expanded for all integer 1 of that equation in terms of vector spherical harmonics. Since $\mathbf{B}^{(3)}$ and its dual, $-i\mathbf{E}^{(3)}/c$, satisfy the four Maxwell equations, then so do these combinations of novel vectorial spherical harmonics. Each value of l for $M = 0$ in Y^L_{011} defines a different *non-zero* component of $\mathbf{B}^{(3)}$ and $i\mathbf{E}^{(3)}$. Therefore the $L = 1$ components in the expansion of $\mathbf{B}^{(3)}$ are dipolar fields. It follows, finally, that many other aspects of standard electromagnetic theory must be augmented to take into account longitudinal fields, for example the discussion on coherency matrices.

CHAPTER 6. CREATION AND ANNIHILATION OF PHOTONS

We have already had occasion to use creation and annihilation operators in previous chapters, and have seen that the novel and fundamental field algebra (4) can be expressed in terms of these operators, the very existence of which means that photons can be created and annihilated. This is a process which satisfies the law of conservation of energy, and its generalizations, usually described with Noether's theorem [54]. The behavior of photons in a light beam must be described statistically, and since photons are objects which arise from a consideration of wave mechanics, these statistics are also expressible in terms of the Heisenberg uncertainty principle. Theoretical considerations such as these lead into the subject of quantum optics, and towards the prediction [90-95] and recent experimental verification [96, 97] of phenomena, such as light squeezing and anti-bunching, *which have no counterparts in classical optics*. In quantum optics there are eigenstates of the photon, for example the number state, denoted $|n\rangle$ and the coherent state, denoted $|\alpha\rangle$, introduced by Glauber in 1963 [98]. There are numerous textbooks available in this area, which is also reviewed and updated constantly in the appropriate literature [99, 100]. In this chapter some characteristics of quantum optics, and creation and annihilation operators, are discussed within a frame of reference built on the emergence of Eqs. (4). In particular, it is shown that $B^{(3)}$ is defined rigorously in quantum optics, as indicated by the considerations given to one photon in Chap. 3. In that chapter, the concept of photon creation and annihilation was not used, and the simplest case of one photon considered, so that statistics were not necessary.

6.1 THE MEANING OF PHOTON CREATION AND ANNIHILATION

The creation operator, denoted $\hat{a}^+(\kappa)$, and the annihilation operator, denoted $\hat{a}(\kappa)$, are basic to the particle interpretation of field theory. The quantization of a scalar field $\phi(x)$, such as that used in Chap. 4, proceeds by Fourier analysis, in terms of the wavevector κ, an analysis which expands the field $\phi(x)$ in a Fourier integral proportional [54]

Chapter 6. Creation and Annihilation of Photons

to the sum of terms

$$\hat{S} = \hat{a}(\kappa)e^{-i\kappa z} + \hat{a}^{+}(\kappa)e^{i\kappa z}, \tag{199}$$

where $\hat{a}$ and $\hat{a}^{+}$ are *operators*. Since this is a quantization procedure, there are commutation relations between the operators $\hat{a}$ and $\hat{a}^{+}$. At two different points x and x', there are two corresponding wavevectors κ and κ', and it can be shown [54] that the desired commutation relations are

$$[\hat{a}_i, \hat{a}_j] = [\hat{a}_i^{+}, \hat{a}_j^{+}] = 0, \quad [\hat{a}_i, \hat{a}_j^{+}] = \delta_{ij}, \tag{200}$$

where δ_{ij} is a Dirac delta function. Therefore, like operators commute, and it can be shown that the operator defined by the product

$$\hat{N}(\kappa) = \hat{a}^{+}(\kappa)\hat{a}(\kappa) \tag{201}$$

also obeys the commutation relation,

$$[\hat{N}(\kappa), \hat{N}(\kappa')] = 0. \tag{202}$$

This means that the eigenstates of the operators $\hat{N}(\kappa)$ and $\hat{N}(\kappa')$ may be used to form a basis, with eigenvalues $n(\kappa)$ defined by the Schrödinger equation [54]

$$\hat{N}(\kappa)|n(\kappa)\rangle = n(\kappa)|n(\kappa)\rangle. \tag{203}$$

The commutation relations

$$[\hat{N}(\kappa), \hat{a}^{+}(\kappa)] = \hat{a}^{+}(\kappa), \quad [\hat{N}(\kappa), \hat{a}(\kappa)] = -\hat{a}(\kappa), \tag{204}$$

show that [54]

$$\hat{N}(\kappa)\hat{a}^{+}(\kappa)|n(\kappa)\rangle = (n(\kappa)+1)\hat{a}^{+}(\kappa)|n(\kappa)\rangle,$$
$$\hat{N}(\kappa)\hat{a}(\kappa)|n(\kappa)\rangle = (n(\kappa)-1)\hat{a}(\kappa)|n(\kappa)\rangle. \tag{205}$$

so that if the state $|n(\kappa)\rangle$ has eigenvalue $n(\kappa)$, the states

Photon Creation and Annihilation

$\hat{a}^+(\kappa)|n(\kappa)\rangle$ and $\hat{a}(\kappa)|n(\kappa)\rangle$ are eigenstates of $\hat{N}(\kappa)$ with eigenvalues $n(\kappa)+1$ and $n(\kappa)-1$ respectively.

This allows the identification [54] of $\hat{N}(\kappa)$ as a *photon number operator*, i.e., the operator for the number of particles with momentum $\hbar\kappa$ and energy $\hbar\kappa c$. Thus $\hat{a}(\kappa)$ is a *photon annihilation operator* and $\hat{a}^+(\kappa)$ is a *photon creation operator*. Particles (photons) which are quanta of this field obey Bose-Einstein statistics [54] and the photons are therefore bosons, there is no restriction on the number of particles that may exist in the same quantum (e.g. momentum) state. This conclusion follows directly from the postulated position momentum commutation relations of the scalar field, as described, for example, by Ryder [54]. Quantization of the Dirac field leads to fermions, with different types of commutation relations (involving anti-commutators [54]). However, another enigma of the photon is that if it can be described with a Dirac equation, as in Chap. 4, then Bose-Einstein and Fermi-Dirac statistics must be interchangeable, as indeed is the case in the most recent thinking on the subject, reviewed by Vigier [72].

For our present purposes, however, it has become clear that there are operators $\hat{a}^+$ and $\hat{a}$ in quantum field theory which increase and decrease the number of photons in a beam, so that that number is not a constant. This is possible, furthermore, without any infringement of the laws of conservation summarized in Noether's theorem.

6.2 QUANTUM CLASSICAL EQUIVALENCE

There exists a quantum mechanical equivalent of any classical quantity, including the electric and magnetic wave fields associated with electromagnetic radiation. Contemporary quantum field theory solves the classical d'Alembert equation in Fourier integral form [54], the latter being once more proportional to an integral over a sum of operators of the type (199). The route taken to quantization of the electromagnetic field represented by A_μ depends on the gauge being used, e.g. the Coulomb or Lorentz gauge, but leads to the concept of photon creation and annihilation operators as for the quantization of the scalar field $\phi(x)$ (Klein-Gordon field [54]). The details of this procedure are well described in a contemporary monograph such as that by Shore [101] and leads to the result that the electric and magnetic fields of electromagnetism become *field operators* directly proportional to the creation and annihilation operators. Following Kielich *et al.* [102] for example, the electric field operator is typically

$$\hat{E}^{(2)} = \left(\frac{\hbar\omega}{\epsilon_0 V}\right)^{\frac{1}{2}} e^{(2)} \hat{a}(0) e^{-i\phi}, \qquad (206)$$

where V is the volume occupied by electromagnetism in the classical picture, and becomes known as the *quantization volume* in quantum field theory. In Eq. (206) $\hat{a}(0)$ is the annihilation operator of a particle, the photon, with momentum $\hbar\kappa$ and in a circular basis rather than Cartesian. As discussed by Kielich et al. [102] the circular basis has a clear advantage over the Cartesian for solutions of equations of motion. In the circular basis denoted (1), (2) and (3), the transverse field vectors are related to the creation and annihilation operators as follows:

$$\hat{E}^{(1)} = \left(\frac{\hbar\omega}{\epsilon_0 V}\right)^{\frac{1}{2}} e^{(1)} \hat{a}^+(0) e^{i\phi}, \qquad \hat{B}^{(1)} = \left(\frac{\mu_0 \hbar\omega}{V}\right)^{\frac{1}{2}} i e^{(1)} \hat{a}^+(0) e^{i\phi},$$
$$(207)$$
$$\hat{E}^{(2)} = \left(\frac{\hbar\omega}{\epsilon_0 V}\right)^{\frac{1}{2}} e^{(2)} \hat{a}(0) e^{-i\phi}, \qquad \hat{B}^{(2)} = -\left(\frac{\mu_0 \hbar\omega}{V}\right)^{\frac{1}{2}} i e^{(2)} \hat{a}(0) e^{-i\phi}.$$

The well known Stokes parameters of the classical theory are therefore described as *bilinear products* of the creation and annihilation operators as described, for example, by Tanaś and Kielich [103],

$$\hat{S}_0 = \left(\frac{\hbar\omega}{\epsilon_0 V}\right) \hat{a}^+(0) \hat{a}(0) \left(e_X^{(1)} e_X^{(2)} + e_Y^{(1)} e_Y^{(2)}\right),$$

$$\hat{S}_1 = \left(\frac{\hbar\omega}{\epsilon_0 V}\right) \hat{a}^+(0) \hat{a}(0) \left(e_X^{(1)} e_X^{(2)} - e_Y^{(1)} e_Y^{(2)}\right),$$
$$(208)$$
$$\hat{S}_2 = \left(\frac{\hbar\omega}{\epsilon_0 V}\right) \hat{a}^+(0) \hat{a}(0) \left(e_X^{(1)} e_Y^{(2)} + e_Y^{(1)} e_X^{(2)}\right),$$

$$\hat{S}_3 = -\left(\frac{\hbar\omega}{\epsilon_0 V}\right) \hat{a}^+(0) \hat{a}(0) i\left(e_X^{(1)} e_Y^{(2)} - e_Y^{(1)} e_X^{(2)}\right),$$

from which it can be seen that the complex conjugate pairs such as $\boldsymbol{E}^{(1)}$ and $\boldsymbol{E}^{(2)}$ or $\boldsymbol{B}^{(1)}$ and $\boldsymbol{B}^{(2)}$ are simply replaced by bilinear operator products.

This is the classical-quantum equivalence rule which allows the straightforward extension of nonlinear optics into nonlinear quantum optics, with the added richness of non-classical effects such as light squeezing.

6.3 LONGITUDINAL AND TIME-LIKE PHOTON OPERATORS; BILINEAR (PHOTON NUMBER) OPERATORS

Having established classical quantum equivalence it becomes necessary in the light of the new classical developments leading to Eq. (4) to construct operators corresponding to the classical, longitudinal, phase free, spin field $\boldsymbol{B}^{(3)}$ and its dual $-i\boldsymbol{E}^{(3)}/c$ in vacuo. It is both convenient and incisive to base this development on the classical Eqs. (157) and (160) of Chap. 5, and to define photon creation and annihilation operators through one to one identities with the unit operators $\hat{e}^{(1)}$, $\hat{e}^{(2)}$ and $\hat{e}^{(3)}$ in the circular basis,

$$\hat{e}^{(1)} := \hat{a}^{(1)}, \qquad \hat{e}^{(2)} := \hat{a}^{(2)}, \qquad \hat{e}^{(3)} := \hat{a}^{(3)}. \qquad (209)$$

We recall from Chap. 5 that the $\hat{e}$ operators are directly proportional to the classical infinitesimal rotation generators $\hat{J}$ in three space-like dimensions. In the four dimensions of space-time these generators are given in the circular basis by Eqs. (161), and in quantum mechanics they are angular momentum operators. The classical $\hat{e}$ operators in three dimensions are matrix representations of the *axial* unit vectors $\boldsymbol{i}$, $\boldsymbol{j}$ and $\boldsymbol{k}$ used to define the *magnetic* components of the electromagnetic field in vacuo in Eq. (157). The corresponding definition in terms of classical $\hat{J}$ operators (infinitesimal rotation generators) is given in Eq. (160).

In Eq. (209) therefore the quantized creation operator $\hat{a}^{(1)}$ has been identified directly with the classical $\hat{e}^{(1)}$ operator, and the annihilation operator $\hat{a}^{(2)}$ with the $\hat{e}^{(2)}$ operator. In this definition, the $\hat{e}^{(3)}$ operator is therefore associated with $\hat{a}^{(3)}$ longitudinal photon operator which has been designated $\hat{a}^{(3)}$, and which is not associated (Eq. (157c)) with the electromagnetic phase. We shall refer to *a* simply as the longitudinal photon operator in vacuo, or "longitudinal operator". In conventional quantum field theory, the longitudinal operators are discarded, along with the time-like photon operator $\hat{a}^{(0)}$, as unphysical, as discussed in

Chapter 6. Creation and Annihilation of Photons

previous chapters. In view of the newly identified Eqs. (4) of classical field theory, this procedure is no longer tenable, and it becomes necessary to include the operators $\hat{a}^{(3)}$ and $\hat{a}^{(0)}$ in quantum field theory *as physically meaningful*.

In quantum mechanics, the classical $\hat{e}$ operators and $\hat{J}$ operators of Chap. 5 become angular momentum operators, so it becomes clear that the operators $\hat{a}^{(1)}$, $\hat{a}^{(2)}$ and $\hat{a}^{(3)}$ are also angular momentum operators in the circular basis of three dimensional space, a result which is consistent with the well known results of angular momentum theory [103] in quantum mechanics which lead to the identification of the operator $\hat{J}^{(2)}$ as a raising operator and $\hat{J}^{(1)}$ as a lowering operator. The creation operator $\hat{a}^{(1)}$ is therefore a lowering operator of angular momentum and the annihilation operator $\hat{a}^{(2)}$ a raising operator (Sec. 6.4).

The definitions (209) have the great advantage of allowing the corresponding definition of the electric and magnetic fields of electromagnetism in vacuo in terms of bilinear products of creation, annihilation and longitudinal operators. Using the commutator relations (156), (158), (162), (163) and (166) leads to the result

$$[\hat{a}^{(1)}, \hat{a}^{(2)}] = -i\hat{a}^{(3)}, \qquad \text{and cyclic permutations} \qquad (210a)$$

$$\hat{B}^{(1)} = B^{(0)}(\hat{a}^{(3)}\hat{a}^{(2)} - \hat{a}^{(2)}\hat{a}^{(3)})e^{i\phi}, \qquad (210b)$$

$$\hat{B}^{(2)} = B^{(0)}(\hat{a}^{(3)}\hat{a}^{(1)} - \hat{a}^{(1)}\hat{a}^{(3)})e^{-i\phi}, \qquad (210c)$$

$$\hat{B}^{(3)} = iB^{(0)}(\hat{a}^{(1)}\hat{a}^{(2)} - \hat{a}^{(2)}\hat{a}^{(1)}), \qquad (210d)$$

for the magnetic fields in Eq. (160). The fields become bilinear products of $\hat{a}$ photon operators, and so are defined in terms of photon number operators. It is important to note from Eqs. (210) that if $\hat{a}^{(3)}$ is discarded as unphysical, the fields $\hat{B}^{(1)}$ and $\hat{B}^{(2)}$, the usual wave fields, become unphysical, an incorrect result. This result of quantum field theory is equivalent to the result in classical field theory that if $B^{(3)}$ is discarded as unphysical in Eqs. (4), all electromagnetism vanishes in vacuo. The infinitesimal rotation operators $\hat{J}$ (which are angular momentum operators in quantum mechanics) are linked to the $\hat{a}$ operators, following the definition (209), through ing the definition (209), through

$$\hat{J}^{(1)} = -i\hat{a}^{(1)}, \quad \hat{J}^{(2)} = i\hat{a}^{(2)}, \quad \hat{J}^{(3)} = -i\hat{a}^{(3)}, \quad (211)$$

so that angular momentum and photon number operators are related in such a way that if the longitudinal operator $\hat{a}^{(3)}$ is discarded as unphysical, $\hat{J}^{(1)}$ and $\hat{J}^{(2)}$ become unphysical, which is again an incorrect result.

These results can also be generalized to the four dimensions of space-time through Eqs. (179) for the four dimensional rotation and boost generators of the Lorentz group. The correct definitions in vacuo of the electric and magnetic fields of electromagnetism are obtained through the scalar operators generated by the following matrix multiplications, which lead to Eqs. (167) (with the notation $\hat{a}_0^{(1)} := \hat{a}_0^{(2)} := \hat{a}^{(0)} := \hat{a}_0^+ := \hat{a}_0$ and $\hat{a}^{(3)} := \hat{a}_Z^+ := \hat{a}_Z^{(1)} := \hat{a}_Z := \hat{a}_Z^{(2)}$ for longitudinal $\hat{a}^{(3)}$ operators), and $\hat{a}^{(1)} = (\hat{a}_X^{(1)} - i\hat{a}_Y^{(1)})/\sqrt{2}$, $\hat{a}^{(2)} = (\hat{a}_X^{(2)} + i\hat{a}_Y^{(2)})/\sqrt{2}$.

$$\hat{J}_X = \begin{bmatrix} \hat{a}_X^{(1)} & \hat{a}_Y^{(1)} & \hat{a}_Z^{(1)} & i\hat{a}_0^{(1)} \end{bmatrix} \begin{bmatrix} 0 & 0 & 0 & 0 \\ 0 & 0 & -i & 0 \\ 0 & i & 0 & 0 \\ 0 & 0 & 0 & 0 \end{bmatrix} \begin{bmatrix} \hat{a}_X^{(2)} \\ \hat{a}_Y^{(2)} \\ \hat{a}_Z^{(2)} \\ i\hat{a}_0^{(2)} \end{bmatrix} = -i\left(\hat{a}_Y^{(1)}\hat{a}_Z^{(2)} - \hat{a}_Z^{(1)}\hat{a}_Y^{(2)}\right),$$

$$(212)$$

$$\hat{J}_Y = -i\left(\hat{a}_Z^{(1)}\hat{a}_X^{(2)} - \hat{a}_X^{(1)}\hat{a}_Z^{(2)}\right), \quad \hat{J}_Z = -i\left(\hat{a}_X^{(1)}\hat{a}_Y^{(2)} - \hat{a}_Y^{(1)}\hat{a}_X^{(2)}\right).$$

Using Eqs. (180) shows that the electric wave and spin fields in vacuo are relativistically correctly defined by bilinear products (photon number operators) which involve physically meaningful $\hat{a}^{(3)}$ and $\hat{a}^{(0)}$ operators in vacuo, as required. Equations (212) also lead back to Eqs. (210) as required.

Similarly,

$$\hat{K}_X = \begin{bmatrix} \hat{a}_X^{(1)} & \hat{a}_Y^{(1)} & \hat{a}_Z^{(1)} & i\hat{a}_0^{(1)} \end{bmatrix} \begin{bmatrix} 0 & 0 & 0 & 1 \\ 0 & 0 & 0 & 0 \\ 0 & i & 0 & 0 \\ -1 & 0 & 0 & 0 \end{bmatrix} \begin{bmatrix} \hat{a}_X^{(2)} \\ \hat{a}_Y^{(2)} \\ \hat{a}_Z^{(2)} \\ i\hat{a}_0^{(2)} \end{bmatrix} = -i\left(\hat{a}_0^{(1)}\hat{a}_X^{(2)} - \hat{a}_X^{(1)}\hat{a}_0^{(2)}\right),$$

$$(213)$$

$$\hat{K}_Y = -i\left(\hat{a}_0^{(1)}\hat{a}_Y^{(2)} - \hat{a}_Y^{(1)}\hat{a}_0^{(2)}\right), \quad \hat{K}_Z = -i\left(\hat{a}_0^{(1)}\hat{a}_Z^{(2)} - \hat{a}_Z^{(1)}\hat{a}_0^{(2)}\right).$$

Chapter 6. Creation and Annihilation of Photons

The physical meaning of the creation and annihilation operators appearing in Eqs. (212) and (213) can be elucidated further by expressing [54] the four dimensional rotation and boost generators in the form

$$\hat{J}_X = -i\left(Y\frac{\partial}{\partial Z} - Z\frac{\partial}{\partial Y}\right), \quad \hat{K}_X = i\left(ct\frac{\partial}{\partial X} + \frac{X}{c}\frac{\partial}{\partial t}\right),$$

$$\hat{J}_Y = -i\left(Z\frac{\partial}{\partial X} - X\frac{\partial}{\partial Z}\right), \quad \hat{K}_Y = i\left(ct\frac{\partial}{\partial Y} + \frac{Y}{c}\frac{\partial}{\partial t}\right) \quad (214)$$

$$\hat{J}_Z = -i\left(X\frac{\partial}{\partial Y} - Y\frac{\partial}{\partial X}\right), \quad \hat{K}_Z = i\left(ct\frac{\partial}{\partial Z} + \frac{Z}{c}\frac{\partial}{\partial t}\right),$$

allowing the following identifications:

$$X := \hat{a}_X^{(1)}, \quad Y := \hat{a}_Y^{(1)}, \quad Z := \hat{a}_Z^{(1)}, \quad ct := -\hat{a}_0^{(1)}$$

$$\frac{\partial}{\partial X} := \hat{a}_X^{(2)}, \quad \frac{\partial}{\partial Y} := \hat{a}_Y^{(2)}, \quad \frac{\partial}{\partial Z} := \hat{a}_Z^{(2)}, \quad \frac{1}{c}\frac{\partial}{\partial t} := \hat{a}_0^{(2)}. \quad (215)$$

The identification of X with the creation operator $\hat{a}_X^{(1)}$ for example, means that X is to be regarded as a unit length operator, and similarly for Y and Z, these identifications being consistent with those in Eqs. (209) of the $\hat{a}$ and $\hat{e}$ operators. The latter are of course unit length operators because they originate in unit vectors $\mathbf{i}$, $\mathbf{j}$ and $\mathbf{k}$. Similarly the differential unit operator $\partial/\partial X$ is identified with the annihilation operator $\hat{a}_X^{(2)}$ and so on. The time-like operator $\hat{a}_0^{(1)}$ is identified with $-ct$, and the operator $\hat{a}_0^{(2)}$ with the differential operator $(1/c)(\partial/\partial t)$. As expected, time, t, appears explicitly in the definition of the time-like operators, and the latter are clearly *physically meaningful*.

To summarize, the ordinary, transverse wave fields of free space electromagnetism, such as $\mathbf{B}^{(1)}$ and $\mathbf{B}^{(2)}$, have been rigorously described in special relativity using bilinear products of $\hat{a}$ operators, products which are photon number operators and which show that $\hat{B}^{(1)}$ and $\hat{B}^{(2)}$ must be described in terms of *the longitudinal operator* $\hat{a}^{(3)}$ as in Eqs. (210b) and (210c). This result shows conclusively that the conventional approach to electrodynamics [4], in which $\hat{a}^{(3)}$ and the time-like $\hat{a}^{(0)}$ are discarded as unphysical is at best incomplete, and at worst incorrect. Similarly, the physically meaningful $\hat{B}^{(3)}$ field is described in terms of bilinear products in Eq. (210d). The three fields $\hat{B}^{(1)}$, $\hat{B}^{(2)}$ and $\hat{B}^{(3)}$ can also be described directly in terms of *single* $\hat{a}$ operators, as in Eqs. (207), a result which follows from the commutator relation (210a). The latter shows that each $\hat{a}$ operator is described in terms of appropriate bilinear products of $\hat{a}$ operators because these are angular momentum operators. These deductions are necessarily based rigorously in the theory of special relativity, and show that the field $\hat{B}^{(3)}$ is a fundamental magnetic field of the photon, *the longitudinal photomagneton*. It is unreasonable to assert that $\hat{B}^{(1)}$ and $\hat{B}^{(2)}$ are physical and fundamental without automatically implying that $\hat{B}^{(3)}$ is also physical and fundamental.

Experimentally, little is known about the nature of the longitudinal photomagneton $\hat{B}^{(3)}$. A key experiment in this context is the inverse Faraday effect, as described in Chap. 1, which is the phenomenon of magnetization by light. There are several independent corroborations in the literature [20-26] of the original experiment by van der Ziel *et al.* [19], but more data would be very useful. The field $\mathbf{B}^{(3)}$ is new to science, but is rigorously based in special relativity and electrodynamics. As argued already, it is consistent with photon mass, and is rigorously non-zero, real and physical. It is dual to the imaginary and unphysical $-i\mathbf{E}^{(3)}/c$ and is consistent with the fundamental laws of conservation of energy, momentum and charge. In this chapter we have shown that it is rigorously defined in contemporary quantum field theory in terms of appropriate photon number operators, a definition which shows conclusively that the ordinary wave fields $\hat{B}^{(1)}$ and $\hat{B}^{(2)}$ also need the *longitudinal* $\hat{a}^{(3)}$ operator. To discard this as unphysical means discarding $\hat{B}^{(1)}$ and $\hat{B}^{(2)}$ as unphysical, an obviously incorrect result which shows that to discard $\hat{B}^{(3)}$ as unphysical means discarding $\hat{B}^{(1)}$ and $\hat{B}^{(2)}$ as unphysical, and conversely, to accept $\hat{B}^{(1)}$ and $\hat{B}^{(2)}$ as physical means accepting $\hat{B}^{(3)}$ as physical.

This is yet another illustration of the enigmatic nature of the photon and concomitant fields, because its conventional description accepts $\hat{B}^{(1)}$ and $\hat{B}^{(2)}$ as physical and discards $\hat{B}^{(3)}$ as unphysical, despite the experimental evidence to the contrary (for example the inverse Faraday effect). The conventional view has held sway for so long and is so influential in contemporary thought that the notion of a physically meaningful $\hat{B}^{(3)}$ in free space needs more experimen-

98   Chapter 6. Creation and Annihilation of Photons

tal data to underpin the self-evident theoretical arguments of these chapters. Equation (6) of Chap. 1 is already conclusive experimental evidence for $\boldsymbol{B}^{(3)}$, and for the observation of the conjugate product $\boldsymbol{B}^{(1)} \times \boldsymbol{B}^{(2)}$. The recent claim by Lakhtakia [103] that the conjugate product is unobservable ignores the existence of experimental data. The obscure claim by Grimes [104] that $\boldsymbol{B}^{(3)}$ exists but is only fortuitously useful is subjective, $\boldsymbol{B}^{(3)}$ is as fundamental in nature as $\boldsymbol{B}^{(1)}$ and $\boldsymbol{B}^{(2)}$. If not, special relativity ceases to be objective, and more generally, *any* theory of natural philosophy ceases to be objective if claims such as those of Lakhtakia and Grimes are to be accepted. For example, it could be asserted arbitrarily that acceleration in Newton's equation is fundamental, but that force is not.

6.4 LIGHT SQUEEZING AND THE PHOTOMAGNETON $\hat{B}^{(3)}$

When there are many photons in a beam of light, such as a coherent, monochromatic laser beam, the Heisenberg uncertainty principle produces quantum effects which are not present classically. One of these is light squeezing, a subject on which there is now a substantial literature, including special issues of journals and reviews [99, 100]. Kielich and Piatek, for example [104], have summarized the properties which characterize a one mode squeezed state as follows. The average value of the annihilation and creation operators are

$$\langle \hat{a} \rangle = \alpha, \quad \langle \hat{a}^+ \rangle = \alpha^*, \qquad (216)$$

where α is defined by the Schrödinger equaton for the Glauber coherent state $|\alpha\rangle$,

$$\hat{a}|\alpha\rangle = \alpha|\alpha\rangle, \quad \langle\alpha|\hat{a}^+ = \langle\alpha|\alpha^*. \qquad (217)$$

These average values do not depend [104] on the squeezing parameter and are not subject to squeezing effects. The average photon number is the expectation value of the operator $\hat{n}$,

$$\langle \hat{n} \rangle = \langle \hat{a}^+\hat{a} \rangle = |\alpha|^2 + \sinh^2 s, \qquad (218)$$

Light Squeezing and the Phtomagneton $\hat{B}^{(3)}$

and the process of squeezing the vacuum results in the second term on the right hand side. The average photon number is therefore affected by the squeezing phenomenon, as is the variances of the quadrature operators and the Heisenberg uncertainty relation between quadrature operators.

In this section effects of light squeezing are discussed on the three fields $\hat{B}^{(1)}$, $\hat{B}^{(2)}$ and $\hat{B}^{(3)}$, and it is shown that the existence of the enigmatic $\hat{B}^{(3)}$ field is compatible with contemporary light squeezing theory [104].

From Eqs. (159) the unit vector operators in the circular basis become angular momentum operators in quantum mechanics. Thus

$$\hat{e}^{(i)}|\alpha^{(i)}\rangle = |\hat{e}^{(i)}||\alpha^{(i)}\rangle, \qquad i = 1, 2, 3, \qquad (219)$$

where the $|\alpha^{(i)}\rangle$ denote *coherent states of photon angular momentum*. The $\hat{a}$ operators defined in Eqs. (209) are also angular momentum operators which create and annihilate photon angular momentum. Thus

$$\hat{a}^{(i)}|\alpha^{(i)}\rangle = |\hat{a}^{(i)}||\alpha^{(i)}\rangle, \qquad i = 1, 2, 3, \qquad (220)$$

are Schrödinger equations for the coherent angular momentum states of photons in a beam of light, to which quantum mechanical angular momentum theory can be applied intact. A particularly useful account of this theory is given in Chap. 6 of Atkins [105], whose notation is linked to ours by

$$\sqrt{2}\,\hat{a}^{(1)} := \hat{I}^-, \qquad \sqrt{2}\,\hat{a}^{(2)} := \hat{I}^+, \qquad \hat{a}^{(3)} := \hat{I}_z, \qquad (221)$$

where $\hat{I}^-$ and $\hat{I}^+$ are angular momentum lowering and raising operators. All the results given by Atkins in his standard text [105] can therefore be transferred to the coherent angular momentum states of photon angular momentum. From the direct proportionality of magnetic field operators $\hat{B}^{(1)}$, $\hat{B}^{(2)}$ and $\hat{B}^{(3)}$ to the $\hat{e}$ operators described in Chap. 5, it becomes clear that the $\hat{B}$ operators are themselves angular momentum operators. The latter operate on the angular momentum coherent states $|\alpha^{(i)}\rangle$ of a beam of many photons. This leads directly to the result

$$\hat{B}^{(3)}|\alpha^{(3)}\rangle = \pm B^{(0)}|\alpha^{(3)}\rangle, \qquad (222)$$

Chapter 6. Creation and Annihilation of Photons

i.e., that the eigenvalues $\pm B^{(0)}$ of the boson operator $\hat{B}^{(3)}$ are non-zero and observable. These eigenvalues are the classical components of the field $\mathbf{B}^{(3)}$, showing again that the conventional assertion that $\mathbf{B}^{(3)}$ is zero is in error, and that the field operator $\hat{B}^{(3)}$ is directly proportional to an angular momentum operator which is also the $\hat{e}^{(3)}$ or $\hat{a}^{(3)}$ operator.

These results apply unchanged to the photon with mass, which is described by the angular momentum theory of bosons as described by Atkins [105], for example, or in a large number of other standard texts on angular momentum theory in quantum mechanics. It therefore becomes clear that the field $\hat{B}^{(3)}$ exists both for the photon with mass and the photon without mass. It is also clear that the operator $\hat{a}^{(1)}$ is an angular momentum lowering operator (denoted $\hat{I}^-$ by Atkins) and $\hat{a}^{(2)}$ is a raising operator ($\hat{I}^+$). It is well known that the $\hat{I}^+$ and $\hat{I}^-$ operators act as follows on an angular momentum eigenfunction denoted in general by $|j, M_j\rangle$,

$$\hat{I}^+|j, M_j\rangle = \hbar c^+|j, M_j+1\rangle, \qquad \hat{I}^-|j, M_j\rangle = \hbar c^-|j, M_j-1\rangle, \qquad (223)$$

where the eigenvalues are given by

$$\hat{c}^+ = (j(j+1) - M(M+1))^{\frac{1}{2}}, \qquad \hat{c}^- = (j(j+1) - M(M-1))^{\frac{1}{2}}. \qquad (224)$$

Following Atkins [105], $\hat{I}^+$ is a raising operator because when it operates on an angular momentum state with Z component $M\hbar$, it generates from it the angular momentum state with the same magnitude but with Z component one unit greater, $(M+1)\hbar$. The operator $\hat{I}^-$ generates the state one unit lower. Therefore $\hat{a}^{(2)}$ and $\hat{a}^{(1)}$ do exactly the same thing, they create and annihilate photon *angular momentum* states, states which are scalar components in the Z axis of the photon angular momentum vector. (Note carefully that because of our definition $\hat{a}^{(1)} := (1/\sqrt{2})(\hat{a}_x - i\hat{a}_y)$ the $\hat{a}^{(1)}$ creation operator is identified with the angular momentum lowering operator $\hat{I}^-$ and vice versa. There is no conceptual difficulty in this because our definition of $\hat{a}^{(1)}$ is equivalent to arbitrarily labelling, or choosing, the direction of angular momentum in space.)

The effect of the $\hat{a}^{(1)}$ and $\hat{a}^{(2)}$ operators can be compared directly with that of creation and annihilation operators of photon number states, described for example by

Light Squeezing and the Phtomagneton $\hat{B}^{(3)}$

Kielich and Piatek [104],

$$\hat{a}|n\rangle = n^{\frac{1}{2}}|n-1\rangle, \qquad \hat{a}^+|n\rangle = (n+1)^{\frac{1}{2}}|n+1\rangle, \qquad (225)$$

for a given j therefore, these number states can be identified with angular momentum M states as in Eq. (223). The initial coherent state $|\alpha\rangle$ can in general be related to the initial number state $|n\rangle$ through a relation such as that given by Tanaś and Kielich [106],

$$|\alpha\rangle = e^{-|\alpha|^2/2} \sum_{n=0}^{\infty} \frac{\alpha^n}{\sqrt{n!}} |n\rangle . \qquad (226)$$

Therefore the operators $\hat{a}^{(1)}$ and $\hat{a}^{(2)}$ create and annihilate angular momentum components in a coherent beam of N photons. These angular momentum components are eigenvalues of coherent eigenstates $|\alpha^{(i)}\rangle$ of photon angular momentum. The $\hat{e}$, $\hat{a}$, $\hat{J}$ and $\hat{B}$ operators are therefore *all* angular momentum operators of quantum theory, and *all* operate on coherent (Glauber) photon states which are angular momentum eigenfunctions in quantum theory. In a beam of N photons therefore all these operators are subject to the Heisenberg uncertainty principle as applied to angular momentum, leading to the conclusion that they are all subject to light squeezing effects. To assert arbitrarily that the operator $\hat{B}^{(3)}$ is unphysical is the same thing as asserting erroneously that one out of three angular momentum operators is unphysical.

The Heisenberg uncertainty principle for a single photon was described in Sec. 3.4 and it applies also to a coherent state of many photons in a light beam. As discussed in Chap. 4, the existence of a non-zero $\hat{B}^{(3)}$ follows directly from the Heisenberg uncertainty principle, for without a non-zero $\hat{B}^{(3)}$, there is no Heisenberg uncertainty. Our introduction of the $\hat{a}$ operators also has the great advantage that the existing knowledge in light squeezing theory can be used to discuss light squeezing effects in photon angular momentum states. Following Tanaś and Kielich, for example [103], the four Stokes operators can be expressed in terms of bilinear products of $\hat{a}$ operator components in elliptical polarization states of a coherent light beam, such as a laser. As shown by these authors, light squeezing effects occur in the $\hat{S}_1$ and $\hat{S}_2$ operators, but the $\hat{S}_0$ and $\hat{S}_3$ operators are constants of motion, and are not affected by squeezing.

The Stokes operator $\hat{S}_3$ of Tanaś and Kielich [103] is described by the same bilinear product of a operators as that defining $\hat{B}^{(3)}$ in Eq. (210d), and the original derivation [9-15] of $\hat{B}^{(3)}$ was based on this fact. The field operator $\hat{B}^{(3)}$ therefore commutes with the Hamiltonian and is a constant of motion, not subject to light squeezing. The fields $\hat{B}^{(1)}$ and $\hat{B}^{(2)}$ however, defined in Eqs., (210b) and (210c), are subject to light squeezing effects because they are not constants of motion. These fields, unlike the Stokes operators $\hat{S}_1$ and $\hat{S}_2$, exist in circular as well as elliptical polarization, and are of course the ordinary transverse magnetic wave fields. *We arrive at the fundamental conclusion that the wave fields of electromagnetism are subject to light squeezing effects when expressed as operators of quantum optics.* These squeezing effects are due to the fact that both $\hat{B}^{(1)}$ and $\hat{B}^{(2)}$ are defined in terms of products such as $\hat{a}^{(3)}\hat{a}^{(2)}$ which are photon number operators. For this conclusion it is essential that $\hat{a}^{(3)}$ and $\hat{a}^{(0)}$ be regarded as physically meaningful as discussed earlier in this chapter. The cyclic operator relations in Eqs. (210) are simply angular momentum commutator relations with the inclusion of exponential phase factors of opposite sign for $\hat{B}^{(1)}$ and $\hat{B}^{(2)}$.

CHAPTER 7. EXPERIMENTAL EVIDENCE FOR $\hat{B}^{(3)}$

Much of the discussion in the foregoing chapters has revolved around the recent discovery of the longitudinal photomagneton $\hat{B}^{(3)}$, because the latter represents a new dimension in electrodynamics, in free space and also in matter. It is therefore necessary to devote a chapter or two to the experimental evidence for $\hat{B}^{(3)}$ and suggestions for further experiments with which to measure its presence. Equation (6) of Chap. 1 shows that the magnetization observed in the inverse Faraday effect (IFE) vanishes if $\boldsymbol{B}^{(3)}$ were zero, showing that $\boldsymbol{B}^{(3)}$ is non-zero experimentally and also showing, as discussed already, that it has all the properties of magnetic flux density. If this were not the case, then $\boldsymbol{M}^{(3)}$ could not be a magnetization, and could not be observable as such. The cyclic symmetry of the classical Eqs. (4), and the isomorphism of this algebra with that of the Lorentz group of special relativity shows that the field $\boldsymbol{B}^{(3)}$ is "as fundamental" as the fields $\boldsymbol{B}^{(1)}$ and $\boldsymbol{B}^{(2)}$. Indeed, the numerous standard texts now available in classical electrodynamics allow for the fact that the Maxwell equations can be solved to give longitudinal magnetic and electric fields in free space. However, these are discarded as being of no interest because wave fields must be transverse. Equations (4) now show that while the wave fields $\boldsymbol{B}^{(1)}$ and $\boldsymbol{B}^{(2)}$ are still transverse, as usual, they are linked inevitably to the spin field $\boldsymbol{B}^{(3)}$, which becomes the photomagneton $\hat{B}^{(3)}$ in quantum theory. The spin field is not a wave field because it has no phase dependence, and in quantum theory is generated directly by, and is directly proportional to, the spin angular momentum whose eigenvalues are $\pm\hbar$ for the photon without mass, or $0, \pm\hbar$ for the photon with mass. In other words the ordinary conjugate product, proportional to the antisymmetric part of light intensity, and denoted $\boldsymbol{B}^{(1)} \times \boldsymbol{B}^{(2)}$, has been identified as $iB^{(0)}\boldsymbol{B}^{(3)}$, where $\boldsymbol{B}^{(3)}$ is a real and physical magnetic flux density. The question that remains to be answered by experiment is whether $\boldsymbol{B}^{(3)}$ can act at first order on matter, e.g. the liquids and solids in which the inverse Faraday effect was first observed [19], or whether it must always act at second order because it is always generated by the second order product $\boldsymbol{B}^{(1)} \times \boldsymbol{B}^{(2)}$. Equations (4) and developments thereof prove without doubt however that: 1) $\boldsymbol{B}^{(3)}$ is non-zero; 2) $\boldsymbol{B}^{(3)}$ is a magnetic flux density, i.e., has all the known properties of magnetic flux density.

Whatever the outcome of these experiments it is already clear that the algebra (4) is isomorphic with the Lie algebra of infinitesimal rotation and boost generators of the Lorentz group, and therefore has the happy consequence of making electrodynamics more self-consistent and less obscure. This finding was illustrated in Chap. 6 using longitudinal and time-like $\hat{a}$ operators, and has important consequences in field theory, not least of which is to point towards the existence of finite photon mass. This may be the most useful consequence of the establishment of Eqs. (4); another is the removal of the obscurities involved in the quantization of the electromagnetic gauge field, specifically, $\hat{a}^{(3)}$ and $\hat{a}^{(0)}$ no longer have to be regarded arbitrarily as "unphysical" in well known procedures such as that of Gupta and Bleuler, described for example by Heitler [107] or Ryder [54]. In the following we review the experimental evidence for $\boldsymbol{B}^{(3)}$ from various sources, and show that considering all sources combined, the existence and observation of $\boldsymbol{B}^{(1)} \times \boldsymbol{B}^{(2)} = iB^{(0)}\boldsymbol{B}^{(3)}$ is beyond reasonable doubt. Some indications are also reviewed of the ability of $\boldsymbol{B}^{(3)}$ to act at first order, but these are not yet unequivocal. Much further experimental work is needed to elucidate the fundamental nature of $\boldsymbol{B}^{(3)}$.

Before proceeding to the description of the experimental evidence for $\boldsymbol{B}^{(3)}$ some recent criticisms are discussed briefly. Barron [53] has suggested that the existence of $\boldsymbol{B}^{(3)}$ would violate $\hat{C}$ symmetry. Chap. 2 however, shows that Eqs. (4), which define $\boldsymbol{B}^{(3)}$, conserve all the discrete symmetries, including $\hat{C}$, and discusses the meaning of $\hat{C}$ in the context of classical electrodynamics. In so doing, it was argued that the amplitude $B^{(0)}$ is to be regarded as positive definite in electrodynamics, an inference which follows from the fact that e is regarded conventionally as negative definite. Barron [53] also asserts that $\boldsymbol{B}^{(3)}$ is zero because it violates $\hat{C}$, but this assertion is clearly untenable because if $\boldsymbol{B}^{(3)}$ were zero, the ordinary conjugate product, $\boldsymbol{B}^{(1)} \times \boldsymbol{B}^{(2)}$, would vanish, a trivially incorrect result. Lakhtakia [103] appears to argue that $\boldsymbol{B}^{(1)} \times \boldsymbol{B}^{(2)}$ is non-zero, and therefore inevitably equal to $iB^{(0)}\boldsymbol{B}^{(3)}$, but is unobservable. This contradicts experimental data from several sources, and therefore Lakhtakia's paper appears to be fundamentally incompatible with experience. It does not refer, for example, to the well known experiments by van der Ziel et al. [19] which first demonstrated the IFE, and to none of the corroborative experimental work which has appeared since then [20-26]. Lakhtakia argues that $B^{(3)}$ exists in elliptical polarization, which is an obvious consequence of the well known fact [106] that elliptical polarization is a weighted mixture of two circular polarizations. (Equations (4) are written in circular polarization.) Grimes [104] appears to argue that $B^{(3)}$ is non-zero but only "fortuitously" useful. This is subjective opinion, $B^{(3)}$ is as useful as any other quantity in electrodynamics. The criticism by Barron [53], who asserts that $B^{(3)}$ is zero, and by Lakhtakia [103] and Grimes [104] who assert that $B^{(3)}$ is *not* zero, but is somehow "not fundamental" are symptomatic of the considerable confusion which can arise if the notion is adhered to that all solutions of Maxwell's equations in free space must be transverse. In attempting to assert subjectively the non-existence of $B^{(3)}$ on this basis, these authors have inadvertently contradicted each other diametrically. In the light of our arguments in previous chapters, these criticisms appear to be superficial. For example, Barron [53] does not refer to the defining Eqs. (4) of the classical theory, and appears to base his argument on diagrams, a procedure which is subjective as argued in Chap. 2. Lakhtakia [103] asserts the non-observability of the ordinary conjugate product $B^{(1)} \times B^{(2)}$ and ignores all experimental evidence to the contrary (see Sec. 7.1). Grimes [104], in an obscure paper, appears to accept that $B^{(3)}$ is non-zero (as is inevitable from Eqs. (4)) but that it can be useful only "fortuitously", a conclusion which is outside the bounds of objective natural philosophy. We stress that Barron [53] comes to precisely the opposite conclusion ($B^{(3)}$ =? 0) to that of Lakhtakia [103], echoed by Grimes [104], ($B^{(3)} \neq 0$) so that in our opinion, these criticisms merely demonstrate a subjective unwillingness to accept Eqs. (4). They are therefore valueless scientifically and have the overall effect of producing confusion in the literature.

7.1 THE INVERSE FARADAY EFFECT, MAGNETIZATION BY LIGHT

The inverse Faraday effect is the phenomenon of magnetization by circularly or elliptically polarized light, and was proposed independently by Piekara and Kielich [108] and by Pershan [109]. It can be shown that the magnetization (ensemble averaged induced magnetic dipole moment) is directly proportional to the conjugate product of light in the form $\boldsymbol{E}^{(1)} \times \boldsymbol{E}^{(2)}$. A recent analysis is given, for example, by Woźniak et al. [26]. In free space we have

$$\boldsymbol{E}^{(1)} \times \boldsymbol{E}^{(2)} = c^2 \boldsymbol{B}^{(1)} \times \boldsymbol{B}^{(2)} \qquad (227)$$

and as shown in Eq. (131), the conjugate product is proportional to the antisymmetric part of light intensity, and is a standard feature in nonlinear optics [15]. It is described clearly by Wagnière [110] in the context of novel birefringence phenomena of light, and is reviewed in magneto-optics by Zawodny [15]. The systematic and pioneering work of Kielich et al. [15] results in expressions such as follows,

$$m_i^{(ind)} = \frac{1}{4}\ ^m\chi_{ijk}^{ee}(0;\omega,-\omega)\, E_j(\omega) E_k^*(\omega), \qquad (228)$$

for the light induced magnetic dipole moment $m_i^{(ind)}$ in the IFE. Here $^m\chi_{ijk}^{ee}(0;\omega,-\omega)$ is a molecular property tensor, a hyperpolarizability. Atkins and Miller [111] and Manakov, Ovsiannikov and Kielich [112] have developed the core expression (228) in quantum field and semi-classical theory. In this section, the IFE is discussed in terms of $\boldsymbol{B}^{(3)}$ with reference to the well known experimental results of van der Ziel et al. [19].

Using the vector $\boldsymbol{B}^{(3)}$ of the classical theory it becomes straightforward [15] to develop any magnetic effect of light, in any polarization. The magnitude of $\boldsymbol{B}^{(3)}$ is given in S.I. units by

$$|\boldsymbol{B}^{(3)}| = B^{(0)} \sim 10^{-7} I_0^{\frac{1}{2}}, \qquad (229)$$

where I_0 is the intensity of the beam in $W\, m^{-2}$. So in, for example, a circularly polarized laser beam of intensity $10^4\ W\, m^{-2}$ ($1.0\ W\, cm^{-2}$) the magnitude of $\boldsymbol{B}^{(3)}$ is about 10^{-5} Tesla, or about $0.1\ G$, roughly a tenth of the earth's magnetic field. It can be shown then that the inverse Faraday effect can be described [14] in samples with a permanent magnetic dipole moment $m_Z^{(0)}$ as

Inverse Faraday Effect, Magnetization by Light

$$M_z \sim \frac{N}{3}c^2\left({}^m\gamma_{123}^{ee} + {}^m\gamma_{231}^{ee} + {}^m\gamma_{312}^{ee}\right)B^{(3)2} \tag{230}$$

$$+ \frac{N}{kT}\left(\langle m_z^{(0)2}\rangle_0 B^{(3)} + \frac{\langle \xi_{zz}^2\rangle_0}{4\mu_0^2} B^{(3)3}\right),$$

where the ${}^m\gamma_{ijk}^{ee}$ denote hyperpolarizability components, N is the number of molecules per m^3; kT the thermal energy per molecule; and where ξ_{ij} is the molecular magnetizability. In samples where $\langle m_z^{(0)2}\rangle_0$ is zero, the term directly proportional to $B^{(3)}$ is zero. These samples include the diamagnetic liquids used by van der Ziel et al. [19]. In samples where there is a permanent net magnetic dipole moment, and if $B^{(3)}$ is capable of acting at first order, there should be observable experimentally a light induced magnetization term proportional to the square root of intensity mixed in with the others of Eq. (230). This term, if observed, would provide unequivocal experimental evidence for the ability of $B^{(3)}$ to act at first order. The data of van der Ziel et al. [19] already provide evidence for the existence of a nonzero $B^{(3)}$ acting at second order.

To estimate [14] the various orders of magnitude of the contributing terms in Eq. (230) the magnetic dipole moment m is set at about a tenth of the Bohr magneton, i.e., at about $10^{-24} J T^{-1}$. A rough order of magnitude estimate of the hyperpolarizability ${}^m\gamma_{ijk}^{ee}$ can be obtained from the Faraday effect theory of Woźniak et al. [108] as about $10^{-45} A m^4 V^{-2}$ for a typical diamagnetic. For a paramagnetic with a permanent magnetic dipole moment it is assumed that the hyperpolarizability may be about 100 times bigger, i.e., $10^{-43} A m^4 V^{-2}$. In Eq. (230) N is typically about 10^{28} molecules m^{-3} and kT at $4 \times 10^{-21} J$ molecule^{-1}, equivalent to $300 K$. A $B^{(3)}$ field of 1.0 Tesla is obtained from a pulse of mode locked, circularly polarized, radiation of intensity about $3 \times 10^{15} W m^{-2}$. Using these figures gives an order of magnitude of magnetization of about $2.5 A m^{-1}$ for the term in $B^{(3)}$, about $2.0 A m^{-1}$ for the term in $B^{(3)3}$, and about $30 A m^{-1}$ for the temperature independent term in $B^{(3)2}$. Clearly, these figures depend on the estimates used above, but all three terms contribute in general.

In their original experiments van der Ziel et al. [19], with a peak intensity of $10^{11} W m^{-2}$ from a pulse of giant ruby laser radiation, observed a magnetization of about $0.01 A m^{-1}$

in 3.1% Eu^{++} doped calcium fluoride glass. They also verified the results with several organic and inorganic liquids and doped glasses at low temperature. The magnetization was detected with a 30 turn coil and spurious signals carefully eliminated. Oscilloscope traces clearly showed signals of opposite sign for opposite senses of circular polarization, and results were cross checked with measurements of the Verdet constant. Graphs of magnetization versus $1/T$ and laser intensity showed a linear dependence (Fig. (2) of Ref. 19), as described in Eq. (6) of Chap. 1. Several different checks were made of the fact that the IFE was indeed being observed, for example the $1/T$ behavior was found to be independent of laser power, and the plots of magnetization versus intensity found to be linear for a variety of liquids and paramagnetic glasses. In the latter, the effect is described by a simple equation of the type (6) of Chap. 1, indicating that $M^{(3)}$ is proportional to $iB^{(0)}B^{(3)}$ multiplied by the imaginary part of a molecular hyperpolarizability as described theoretically by Woźniak et al. [26].

The recent claim made by Lakhtakia [103] that $iB^{(0)}B^{(3)} = B^{(1)} \times B^{(2)}$ is unobservable ignores the careful cross checking procedures carried out by van der Ziel et al. [19] and indeed ignores all the available experimental data on magnetization by light [15].

Finally in this section, it is noted that van der Ziel et al. [19] refer to the term in the conjugate product $E^{(1)} \times E^{(2)}$ in the paramagnetic IFE as an effective magnetic field term, and it is now clear that the magnetization is caused by a non-zero $B^{(3)}$, a real and *physical* magnetic field. There is no sign in the data of van der Ziel et al. [19] that this acts at first order, but it is clear that it acts at second order through Eq. (6), thus proving its relation to $B^{(1)} \times B^{(2)}$ in Eqs. (4).

7.2 OPTICAL NMR, FIRST ORDER EFFECT OF $B^{(3)}$

The phenomenon of magnetization by light has been used recently [113] to shift NMR resonances with circularly polarized light far from optical resonance. It can be shown [114, 115] that the observed shifts are about *fourteen orders of magnitude* larger than expected from conventional perturbation theory applied to shielding constants in NMR, constants which determine the chemical shift used to identify the sample analytically. The observed shifts must therefore be due to a hitherto unrecognized magnetic property of light, a property which does not time average to zero, and which affects the sample at first order in the magnetic field of electromagnetism. It is tempting to assert that this property is the field $\boldsymbol{B}^{(3)}$, but the experimental evidence is not yet unequivocal. In this section we describe briefly the results of the first ONMR experiment [113, 116] and the extent to which the simplest, (vacuum $\boldsymbol{B}^{(3)}$) theory can account for the data.

If the experimental results of Warren et al. (113, 116) are accepted, however, they indicate without reasonable doubt that there exists a magnetic property of light which has not been recognized hitherto. The reasons for this deduction are discussed as follows.

The ONMR technique relies in the simplest case on the use of a continuous wave, circularly polarized laser of low intensity directed into the spinning sample tube of a conventional NMR spectrometer. (The same technique can be used, in principle, for OESR.) The first series of ONMR experiments, carried out by Warren and co-workers [113, 116, 117], involved several painstaking checks of repeatability and reproducibility, and were carried out under a variety of conditions with several samples. Special care was taken to remove heating artifacts, or otherwise to differentiate the effect of heating from the magnetic effect of laser light. A conclusive demonstration [116] involves heating the sample to equilibrium with the laser, then rapidly switching from left to right circular polarization, whereupon the resonance line shifts upwards or downwards in frequency in synchronization. Such rapid modulations cannot be due to heating, and the observed shifts are many orders of magnitude greater than allowed for in standard perturbation theory applied [114, 115] at first order in laser intensity (second order in the field).

A detailed discussion of the application of $\boldsymbol{B}^{(3)}$ theory to these results has been given elsewhere [15] and shows that if the vacuum value of $\boldsymbol{B}^{(3)}$ is used, the observed shifts appear to be *overestimated* theoretically by about one or two orders of magnitude. This compares with an *underestimation* by the conventional second order perturbation theory [114, 115] of about fourteen orders. In this rough and ready understanding therefore, the $\boldsymbol{B}^{(3)}$ theory does far better than the second order theory but does not yet lead to a satisfactory description of the data [113, 116, 117], the most significant feature of which is the fact that reversing the circular polarization reverses the direction of the light induced shift.

This property is reminiscent of the similar observation in the inverse Faraday effect [19] that reversing the direction of circular polarization reverses the sign of the voltage produced by the induction coil. These are raw pieces of data which point towards the ability of light to magnetize, at first order in ONMR and at second order in the IFE. The ONMR data, however, display [113, 116, 117] what appears to be a complicated dependence on the light intensity, and refinement of the prototype technique is necessary. This may become possible in future with improvements in instrumentation. The reversal of shift direction with laser polarization is however a conclusive piece of evidence for a $\boldsymbol{B}^{(3)}$ mechanism at first order, *because $\boldsymbol{B}^{(3)}$ also reverses sign with circular polarization and does not time average to zero as do $\boldsymbol{B}^{(1)}$ and $\boldsymbol{B}^{(2)}$ when acting at first order.* If attention is focused on this feature, it becomes clear that there can be no other mechanism for ONMR shifts which reverse direction with circular polarization. This is simply because at the optical frequencies used in the experiment [113, 116, 117], any *first* order mechanism due to the oscillatory $\boldsymbol{B}^{(1)}$ and $\boldsymbol{B}^{(2)}$ would average to zero because the phase is $e^{i\phi} = \cos\phi + i\sin\phi$, whose average value is zero. Any *second* order mechanism due to $\boldsymbol{B}^{(1)}$ and $\boldsymbol{B}^{(2)}$ is fourteen orders of magnitude too small in standard second order perturbation theory as worked out in Ref. 114. If we are not to dismiss the whole series of experiments [113, 116, 117] as producing mere artifact, we are driven inevitably to the conclusion that the shifts are due to a $\boldsymbol{B}^{(3)}$ mechanism, because there is no other reasonable possibility. (The fact that perturbations at first order in the transverse fields of light average to zero is the very reason why the perturbation theory is always conventionally applied at second order in the transverse field, i.e., at first order in the light intensity.)

If the shifts are caused by a $\boldsymbol{B}^{(3)}$ mechanism, however, then there must be a reason why they are much smaller than expected on the basis of a vacuum $|\boldsymbol{B}^{(3)}|$ of about $10^{-7} I_0^{1/2}$ Tesla. This is not clear at present, possibly there is a simple experimental reason, for example loss of intensity in guiding the laser into the sample tube. Another possibility is that the vacuum value of $|\boldsymbol{B}^{(3)}|$ (a property of light) does not reach the nucleus because of the surrounding electrons. This would tend to fit in with the observation [113, 116, 117] that the shifts are (usefully) site specific, i.e., dependent on the electronic environment of the nucleus. Obviously, much more work is needed to clear up these features, but the reversal of shift with circular polarization is conclusive, and very unlikely to be an artifact of experiment.

It is therefore concluded that ONMR is evidence, albeit incompletely understood, of the ability of $B^{(3)}$ to act at first order at the nucleus.

7.3 THE OPTICAL FARADAY EFFECT (OFE)

We refer to the optical Faraday effect (OFE) as the rotation of the plane of polarization of a linearly polarized probe beam by a second, circularly polarized, pump laser. The latter substitutes for the magnetic field of the ordinary Faraday effect [15]. The circularly (or elliptically) polarized pump laser generates a $B^{(3)}$ field which acts at first and second order. This means that the field $B^{(3)}$ can shift the ordinary magnetic circular dichroism (MCD), for example, set up in the sample by a standard static magnetic field. The complete MCD spectrum is shifted upwards or downwards in frequency according to the sense of circular polarization of the pump laser applied to the sample in the MCD spectrometer. If $B^{(3)}$ acts at first order its effect is simply to augment or decrease the effective magnetic field applied to the sample,

$$B = B_0 \pm B^{(3)}, \qquad (231)$$

where B_0 is the static magnetic flux density used in the spectrometer. Sanford et al. [22] have recently reported this effect in ferroelectric samples such as $CdCr_2Se_4$. These authors noted a large shift in the Faraday rotation spectrum of this compound at 78 K, a shift caused by a circularly polarized laser. It was observed [22] that the shift was reversed by reversing the circular polarization of the pump, the displacement being to lower energy when right circularly polarized light was used and vice versa. The intensity of the pump laser was only $0.7\,W\,cm^{-2}$, and the sample thickness was only 25 μm, *so the OFE effect observed by Sanford et al.* [22] *is a very large one.* It was observed at 78 K that the complete Faraday rotation spectrum near 1000 nm was shifted without distortion of the bandshape to higher or lower frequencies by the pump laser, an effect which was interpreted in terms of a pump induced increment in the effective magnetic field, which was observed at remarkably low pumping levels, and which was reported to be ten times greater in magnitude in circular than in linear polarization.

The classical $B^{(3)}$ can be used straightforwardly to explain these observations qualitatively, because as in Sec. 7.1 and 7.2, $B^{(3)}$ changes sign with circular polarization and shifts the MCD spectrum of the ferromagnetic upwards or downwards in frequency without distorting the bandshape, precisely as observed. The field $B^{(3)}$ is, as we have argued, the expectation value of the free space photomagneton $\hat{B}^{(3)}$, generated directly by photon spin. The operator $\hat{B}^{(3)}$ is therefore a phase free spin operator, generated directly by the photon's non-zero intrinsic angular momentum. If it is assumed a priori that the classical $B^{(3)}$ acts directly at first order on the electrons of the sample in the OFE, its overall effect, as we have argued, would be simply to shift the MCD spectrum in either direction, depending on the laser's sense of circular polarization. If the pump laser were linearly polarized there would be no shift, an inference which is again borne out qualitatively by the data of Sanford et al. [22] in their Fig. (2). These authors report that circularly polarized pumping was an order of magnitude more effective than linearly polarized pumping.

Assuming in the first approximation that the vacuum value of $B^{(3)}$ can be used, standard Faraday effect theory [15] gives the following expression for the angle of rotation of a linearly polarized probe beam caused by $B^{(3)}$ of the pump beam,

$$\Delta\theta \sim \frac{1}{12}\omega\mu_0 c l N |B^{(3)}| \epsilon_{\alpha\beta\gamma} \left({}^e\gamma^{em}_{\alpha\beta\gamma}(f) + \frac{m_{n\alpha}}{kT}\alpha''_{\beta\gamma}(f) \right), \qquad (232)$$

where ω is the angular frequency of the probe, μ_0 the free space permeability, c the speed of light in vacuo, l the sample thickness, N the number of atoms or molecules per cubic meter of sample, $\epsilon_{\alpha\beta\gamma}$ the Levi-Civita symbol, kT the thermal energy per molecule, and ${}^e\gamma^{em}_{\alpha\beta\gamma}(f)$, $m_{n\alpha}$ and $\alpha''_{\beta\gamma}(f)$ the standard Faraday effect molecular property tensors [15], properties of the sample. Respectively, these are the hyperpolarizability, permanent magnetic dipole moment (if non-zero), and antisymmetric polarizability. The semi-classical structure of the hyperpolarizability is given by Woźniak et al. [26], and that of the other two property tensors is standard [15].

It is clear from Eq. (232) that $\Delta\theta$ is directly propor-

The Optical Faraday Effect (OFE)

tional to $B^{(3)}$ and will be shifted without distortion upwards or downwards in frequency according to the sense of circular polarization. Such a shift should be proportional experimentally to the square root of the pump laser intensity. No intensity dependence of the shift is reported by Sanford et al., but in other respects their results are explained qualitatively by Eq. (232). Sanford et al. themselves interpret the shift in terms of $\pm(3/2)J'f$ where J' is the first order intra-Cr exchange and f the fraction of chromium sites with excited electrons. In the notation of Eq. (232) this mechanism is described semi-classically through the term $m_{n\alpha}\alpha^{//}_{\beta\gamma}$, which is temperature dependent. The detailed mechanism of Sanford et al. [22] is contained within the structure of these molecular property tensors. Sanford et al. note [22] that as the temperature is changed from 78 K, where the Faraday rotation peaks are close to the fixed pump frequency, there is a gradual loss of symmetry in shifts for the right and left circularly polarized pump. Sanford et al. [22] explain this gradual loss of symmetry as being caused by the temperature dependence of the energy splitting between sub-bands, a mechanism which is allowed for in Eq. (232) through the fact that the optical resonance structure for the molecular property tensors depends on the energy splitting between sub-levels and on temperature. For example, the resonance structure of the IFE, worked out for a model two level atom by Woźniak et al. [26], shows asymmetry near and away from optical resonance. Sanford et al. [22] refer to their interpretation of their data as a first order magnetic coupling, and Eq. (232) is, significantly, one to first order in the magnetic field $B^{(3)}$.

It is important in further work to study these effects as a function of pump laser intensity, in order to separate out the effects of $B^{(3)}$ at first and higher orders. If $B^{(3)}$ were zero, however, as suggested by Barron [53], there would be no effects of the type described in this chapter, at any order in $B^{(3)}$, something which is obviously contrary to experience in the three independent experiments of Sec. 7.1 to 7.3. Therefore the spin field $B^{(3)}$ is not zero, and effects due to it have been observed unequivocally at second order in the IFE (Sec. 7.1). By inference, all the well known magneto-optic effects reviewed recently by Zawodny [15] are due to $B^{(3)}$ and its various interactions with material matter. There are signs that $B^{(3)}$ can act at first order, for example in ONMR (Sec. 7.2.), and it is interesting to note that the nonlinear Faraday rotations reported recently by Frey et al. [118] fall on a straight line when plotted against the *square root* of pump intensity. These effects are discussed more fully elsewhere [15].

From Eq. (232) and the experimental results of Sanford et al. [22] a pump of $0.7\ W\ cm^{-2}$ produces at a fixed frequency an ordinate shift of about 10 degrees (roughly 0.2 rad) in the MCD spectrum at 1000 nm (roughly $6\pi \times 10^{14}$ rad s^{-1}), at $78\ K$ and for a sample thickness of 25 microns. Assuming that there are of the order 10^{26} atoms m^{-3} in the sample, we obtain rough estimates of the effective orders of magnitude of $m_{n\alpha}\alpha''_{\beta\gamma}$ and $^e\gamma^{em}_{\alpha\beta\gamma}$ in Eq. (232),

$$|\ ^e\gamma^{em}_{\alpha\beta\gamma}(f)\ | \sim 10^{-34}\ Am^4\ V^{-2}, \qquad |m_{n\alpha}\alpha''_{\beta\gamma}| \sim 10^{-53}\ C^2\ m^2\ T^{-1}. \qquad (233)$$

In a diamagnetic liquid such as CS_2, Woźniak et al. [26] have calculated a value for the hyperpolarizability of the ordinary Faraday effect: $10^{-45}\ Am^4\ V^{-2}$. In a paramagnetic with a permanent molecular dipole moment, the term $|m_{n\alpha}\alpha''_{\beta\gamma}|$ is very roughly of the order $10^{-69}\ C^2 m^2 T^{-1}$. Consistently, therefore, the atomic molecular property tensors in the ferromagnetic sample used by Sanford et al. [22] are much greater in magnitude than in a paramagnetic or diamagnetic. This is experimentally consistent with the fact that some magnetic semiconductors, such as those used by Frey et al. [118] show giant Zeeman and Faraday effects orders of magnitude greater than encountered in ordinary diamagnetics such as water. Sanford et al. [22] have elegantly amplified the effect further by tuning the pump laser to peaks in the original Faraday effect spectrum of the sample. This is precisely what is proposed theoretically by Woźniak et al. [26] for the IFE.

In general, $\boldsymbol{B}^{(3)}$ can also act at second order in the OFE,

$$\Delta\theta \sim \frac{1}{2}\omega\mu_0 c^3 lNB^{(3)2}\left(\langle\alpha^{(II)}_{1XYZ}(f)\rangle + \frac{1}{kT}\langle\alpha''_{1Zn}\alpha''_{1XY}(f)\rangle\right), \qquad (234)$$

here, the angular brackets denote averaging over appropriate molecular property tensors, defined in Ref. 15. This effect depends on the conjugate product directly, expressed as $iB^{(0)}\boldsymbol{B}^{(3)}$, and as such also changes sign between circular polarizations of the pump laser. However, the effect of $\boldsymbol{B}^{(3)}$ at second order is proportional to pump laser intensity, and in general accompanies the effect to first order in $\boldsymbol{B}^{(3)}$ just described.

The Optical Faraday Effect (OFE) 115

7.4 SURVEY OF DATA

As we have seen, there exist signs from the three sources described in this chapter that $B^{(3)}$ can act both at first and second order in material matter. (It can, of course, act also at higher orders under the right conditions.) These three experiments have been chosen to illustrate the experimental existence of $B^{(3)}$ because they were carried out with particular care, and produced incisive results. Mention was also made of the nonlinear optical rotations reported by Frey et al. [118], and of corroborative experiments on the IFE [20-26]. A range of magneto-optic effects, obtained by numerous groups over the past thirty years or more is reviewed by Zawodny [15]. It is well accepted that these nonlinear effects are due to the conjugate product $B^{(1)} \times B^{(2)}$, which can now be expressed as in Eqs. (4) as $iB^{(0)}B^{(3)}$, and therefore in terms of the real field $B^{(3)}$. This simple re-expression of the ordinary conjugate product was not realized until about 1992 [9-15], so that the existence of a field such as $B^{(3)}$ was not suspected. Accordingly, these experimental data were naturally interpreted in the conventional view that all magneto-optic effects of this type must be due to the conjugate product and must therefore be proportional to light intensity. This effect is now recognizable as that of $iB^{(0)}B^{(3)}$ but there is also expected an effect due to $B^{(3)}$ itself, which should be proportional to the square root of light intensity. Because $B^{(1)}$, $B^{(2)}$ and $B^{(3)}$ form a group which is isomorphic with the group of rotation generators, either in space or space-time, the three fields are physical and it is not valid to assert that any one is unphysical. Equations (4) therefore represent a Lie algebra which automatically ensures that $B^{(3)}$ is phase free, having been formed from the product of $B^{(1)}$ and $B^{(2)}$. This product is the ordinary conjugate product of nonlinear optics. A subjective refusal to accept Eqs. (4), such as evidenced in the papers by Barron [53], Lakhtakia [103], and Grimes [104] will lead inevitably to obscure and incorrect results, internal contradictions and qualitative disagreement with experimental data. For example Barron asserts that $B^{(1)} \times B^{(2)}$ exists but that $B^{(3)}$ does not, and this is trivially inconsistent by Eq. (4a). Lakhtakia asserts that the conjugate product is unobservable, which is trivially inconsistent with the data presented in this chapter and reviewed in depth in Ref. 15. Grimes asserts that $B^{(3)}$ is useful by coincidence, whereas Eqs. (4) are algebraic.

CHAPTER 8. THE CONCEPT OF PHOTON MASS

8.1 THE PROBLEM

The idea that the photon may have a non-zero mass was developed by de Broglie over many years of investigation. His first massive photon equations were proposed [119] in 1934, shortly after the emergence of the Proca field equation (68) in 1930 [120]. The work of de Broglie in this area is recorded in numerous books and articles which are accessible through Library of Congress Listings [121]. The de Broglie photon equation of 1934 (not to be confused with the de Broglie Guiding theorem (1)) is described by Goldhaber and Nieto [85] for example as coming from the product of Dirac particle and antiparticle spaces. From the very outset the photon is regarded as a particle with mass, and the anti-photon as an anti-particle with mass. The de Broglie photon equation is related to the Duffin, Kemmer and Petiau equations (similar to Eq. (125)) in their fundamental, reducible, representation. The latter can be defined as a symmetric product, or composite, of two Dirac particle spaces. The Duffin, Kemmer, and Petiau equations reduce [122] to the Klein-Gordon and Proca equations if it is assumed that the wave function transforms as a product of two Dirac wave functions. The Klein-Gordon equation becomes the irreducible representation corresponding to a five dimensional pseudo-scalar equation obtained for a plane wave and diagonal matrix elements. The Proca equation becomes the irreducible representation corresponding to a ten dimensional spin one equation. The de Broglie photon equation on the other hand decomposes into a one dimensional pseudo-scalar irreducible representation; a five dimensional irreducible representation corresponding to a scalar Klein-Gordon equation and a ten dimensional axial vector representation of the Proca equation. Duffin [123] has given a general discussion of the characteristic matrices of covariant, quantum relativistic systems.

The de Broglie massive photon equation of 1934 considers Dirac spaces for particles and anti-particles which are photons and anti-photons (Chap. 2) with mass. The Duffin, Kemmer and Petiau equations [85] on the other hand consider only Dirac spaces, and therefore only photons, which are considered conventionally as their own anti-particles (but see Chap. 2). If these equations are applied to photons (and anti-photons) in the classical limit $m_0 \to 0$, the Maxwell equations of the classical electromagnetic field must be recovered. An example of the Maxwell equations written in the form of a Dirac equation has been given in Eq. (126), following Barut [46]. As seen in Eqs. (83), the d'Alembert, Dirac and Schrödinger equations for the photon without mass become identical in structure, each being represented by a d'Alembertian operating on a potential four-vector A_μ, a Dirac spinor ψ, or a scalar field ϕ.

As discussed in Chap. 2, the $\hat{C}$ operator applied to the photon produces a distinct anti-photon, and this is allowed for in the de Broglie photon equation. The photon in this picture is not its own anti-particle, $\hat{C}$ has the effect of reversing the sign of all four components of the potential four-vector A_μ and of the scalar quantities e, $E^{(0)}$ and $B^{(0)}$. (In contemporary field theory the vacuum is the Dirac sea, and $\hat{C}$ has the effect of changing the particles of the Dirac sea to anti-particles.) Its effect on the de Broglie photon equation is to produce the de Broglie anti-photon equation. The fact that

$$\hat{C}(A_\mu) = -A_\mu \qquad (235)$$

means that the photon is an eigenstate of $\hat{C} = -1$, which in contemporary understanding [52], is a consequence of the covariance of the U(1) electromagnetic sector field equations under $\hat{C}$ and $\hat{C}\hat{P}\hat{T}$ (Chap. 2). The de Broglie, Duffin, Kemmer and Petiau equations, being physical laws, must be invariant under the discrete symmetries, unless there is symmetry violation, a property that implies the conservation of the negative charge parity of the photon (and anti-photon). If negative charge parity is conserved, there is no observable $\hat{C}$ violation in the natural universe, which has evolved (Chap. 2) in such a way that it appears to be composed, naturally, of particles and fields rather than anti-particles and anti-fields. Thus, the fact that all electrons in the natural universe appear to be negatively charged does not mean $\hat{C}$ violation, but is a consequence of initial (boundary) conditions. Positrons and other anti-particles can be manufactured in the laboratory, but appear not to occur naturally. Similarly, some natural products such as sugars and alkaloids occur as one enantiomer, and not as a racemic mixture. This does not imply $\hat{P}$ violation, because the

The Problem

opposite enantiomer can be synthesized. Similarly, the anti-photon can be synthesized under appropriate conditions, although this remains to be demonstrated experimentally.

In contemporary field theory, however, the notion that the photon is its own anti-photon is prevalent, but as discussed in Chap. 2, is apt only for the photon as particle, not for the photon as particle and concomitant field. In this framework the Duffin, Kemmer or Petiau equations are descriptions of the electromagnetic field as a massive gauge field, whose quantization produces well defined massive photons with three space-like polarizations, not two, as in the conventional picture. This equation must be invariant under local U(1) gauge transformations [54]. This leads to the limiting gauge condition described by $A_\mu A_\mu = 0$, which is a consequence of the Dirac condition for vanishingly small photon radius. This condition therefore is valid for finite photon mass (m_0) and an infinitesimally small photon radius, represented relativistically [72] by the radius four-vector r_μ. The four-vector r_μ is orthogonal to the energy momentum vector p_μ of the photon in its rest frame.

The usual contemporary description of the U(1) sector differs from this in that the photon mass is considered to be *identically* zero. Goldhaber and Nieto [85] have shown that there is no evidence for this assertion, and standard tables now show the mass of the photon as non-zero. There can be no logical evidence for identically zero photon mass, because this would imply an infinite range for the electromagnetic field in a universe with presumably finite radius. In contemporary unified field theory, such as SU(5) or GWS, finite m_0 can be made compatible with the powerful results of GWS, for example its prediction of massive intermediate vector bosons. Huang discusses the fact that non-zero m_0 implies a finite electron lifetime within a given model structure.

In contemporary thought, electromagnetic field theory is the U(1) sector of grand unified theory, and it is essential that meaningful consideration be given to the existence of a non-zero photon mass, however powerful the results of GWS or SU(5) may be. The alternative is to accept by assertion an identically zero m_0, for which there can be no experimental evidence as mentioned already. The logic of de Broglie's work therefore remains just as valid as ever. Considerations of non-zero m_0 date from Cavendish's time, and were involved, for example, in Einstein's development of general relativity, in which gravitation bends light as if it had mass. As soon as the possibility is accepted that $m_0 \neq 0$ contemporary theories of gauge invariance are affected at a fundamental level as discussed in Chap. (4). Gauge freedom is lost, the Lorentz gauge always applies, and as discussed in Chap. (1), the d'Alembert equation becomes the Proca equation, an eigenvalue equation. The Coulomb gauge for $m_0 = 0$, identically, is inconsistent with $m_0 \neq 0$ because the Lorentz gauge applies. If (in S.I. units), ϵ_0 being the permittivity in vacuo:

$$A_\mu \equiv \epsilon_0 \left(\mathbf{A}, \frac{i\phi}{c} \right), \qquad (236)$$

as usual [4], Eq. (145) implies the light-like result:

$$\phi = c|\mathbf{A}|, \qquad (237)$$

with the important consequence that A_μ becomes a *physically meaningful* four-vector, as observed in the Aharonov-Bohm effect [15]. This means that all four components of A_μ must be physically meaningful, as argued in previous chapters. Finite photon mass means that the basic idea of the Gupta-Bleuler field quantization procedure [54] must be abandoned, because the longitudinal and time-like components of A_μ (expressed as creation and annihilation operators) *cannot be discarded as unphysical*. As shown in Eqs. (210) of Chap. 6, these *same* longitudinal and time-like components occur in the fully covariant definition of the ordinary *transverse* fields of electromagnetism in free space, the wave fields $\mathbf{B}^{(1)}$ and $\mathbf{B}^{(2)}$. In this view, the spin field $\mathbf{B}^{(3)}$ is described through the X and Y components of creation and annihilation operators, operators which *also* occur in components of a physically meaningful A_μ.

Standing back from received wisdom, it becomes unacceptable to assert that two out of four components of a physically meaningful A_μ must be physically meaningless (i.e., unphysical). This unacceptability of the conventional viewpoint is precisely the conclusion reached on the grounds that m_0 is not necessarily identically zero. Finite photon mass leads to the inference of a non-zero $\mathbf{B}^{(3)}$.

8.2 BRIEF REVIEW OF EXPERIMENTAL EVIDENCE COMPATIBLE WITH $m_0 \neq 0$ AND $B^{(3)} \neq 0$

There can be no experimental evidence for $m_0 = 0$ because the radius of the universe is apparently finite: $m_0 = 0$

Experimental Evidence Compatible with $m_0 \neq 0$ and $B^{(3)} \neq 0$

identically implies an infinite range for electromagnetic radiation, and if the extent of the universe is finite, light received by an earthbound observer from even the most distant galaxies has still travelled a finite distance. The radius of the universe can be estimated experimentally only through an interpretation of the properties of this type of radiation. The hypothesis $m_0 = 0$ therefore has no experimental support, which implies that if the photon be regarded as a particle, it can take finite mass, and can travel at speeds which are different from c. Therefore, the latter becomes a postulated fundamental constant of special relativity, no longer to be identified with the speed of light, whose properties become different in different Lorentz frames, even in free space. The photon with finite mass can be assigned a rest frame, and its properties in any other frame are determined by the Lorentz transformation of special relativity. From the de Broglie Guiding theorem, light of different frequencies propagates at different speeds through a vacuum, an inference which can be tested experimentally as described for example by Goldhaber and Nieto [85], who cover the field up to about 1970. A more recent review by Vigier [72] covers the experimental evidence compatible with a finite m_0 up to about 1992. It is clear that the twenty or so years separating these two review articles have seen a great increase in interest in the inference that m_0 is not identically zero. Unfortunately, contemporary gauge theory is still dominated by the assertion that $m_0 = 0$ identically; and there is an obvious incompatibility between these two branches of physics. The theory of finite photon mass attempts to bridge this gap while retaining the powerful results of unified field theory, for example GWS, SU(5), and chromodynamics, without having to assert $m_0 = 0$. Experimental work on these bridging theories is much needed [72].

In a brief overview, the various types of contemporary evidence cited by Vigier [72] are as follows. There is a direction dependent anisotropy of light in the apex [124] of the 2.7 K background of microwave radiation in the universe, an anisotropy which is compatible with finite photon mass. Experiments on the existence of superluminal action at a distance [125] have been performed and are being repeated with increasing accuracy with the overall intention of proving the central idea of non-locality in the quantum potential [126]. Other types of contemporary experiments investigate directly the Heisenberg uncertainty principle for single photons, because this is central to the interpretation of quantum mechanics, through the Copenhagen or Einstein-de Broglie viewpoints. Interesting recent discussions include those of Selleri [127] and Grigolini [128]. Questions of non-locality, simultaneous existence of wave and particle, and finite m_0, are interrelated inextricably so that a complex of fundamental ideas is being subjected to experimental investigation at the time of writing. The existence of photon and particle like trajectories is being tested experimentally in optical and neutron self interference experiments. This question is familiar as the Einweg-Welcherweg problem, which has persisted throughout the twentieth century. The fundamental ideas of the Copenhagen School, originally proposed by Bohr and others, in which light is considered to be made up of particles and waves of probability, and that of the Einstein-de Broglie School, in which light is made up simultaneously of particles and of waves, both of which are observable simultaneously, are being tested directly. The physical co-existence of wave and particle is therefore a central point of interpretation: it is possible in the Einstein-de Broglie interpretation, impossible in the Copenhagen interpretation.

Experimental tests for the existence of the quantum potential, which is responsible for the piloting of photons by wave or spin fields, have been devised [129] using coherent intersecting laser beams. This type of evidence for the Einstein-de Broglie interpretation of dualism is also provided by an experiment such as that of Bartlett and Corle [129] which measures the Maxwell displacement current in vacuo and without electrons. Evidence from such sources is augmented by laser induced fringe patterns, showing an observable enhancement of photon energy due to the quantum potential. In this context, the optical equivalent of the Aharonov-Bohm effect using the $\boldsymbol{B}^{(3)}$ field would be a critical test of the ability of $\boldsymbol{B}^{(3)}$ to act at first order. Experiments such as those of de Martini *et al.*, discussed recently by Vigier [72] show that it is possible to pass continuously from Bose-Einstein to Maxwell-Boltzmann (Poisson) statistics in an ensemble of photons. The passage from one type of statistics to the other can be explained in terms of non-locality in the quantum potential, which results in non-local action at a distance, currently a critical question in quantum mechanics.

In astrophysics, the consequences of non-zero photon mass are many and varied, and have been considered repeatedly throughout the twentieth century. Foremost among these is that the Proca equation, as we have argued, produces longitudinal photons which do not affect the validity of the Planck radiation law. The field $\boldsymbol{B}^{(3)}$, as we have argued, is for all practical purposes identical in the d'Alembert and Proca equations, and longitudinal photons play a role in the definition of $B^{(1)}$, $B^{(2)}$ and $B^{(3)}$, all three fields being physical and all producing observable effects in material matter such as magnetization at first and higher orders as described in Chap. 7. The existence of $B^{(3)}$ is furthermore implied by finite photon mass, as we have argued already. In the Proca equation, the field $B^{(3)}$ decreases exponentially, and over large enough distances, Z, on a cosmic scale (e.g. light from distant galaxies), the decrease in $B^{(3)}$ might become observable. This would be a direct measurement of photon mass through the parameter $\xi := m_0 c/\hbar$ (Eq. (9b)).

The photon flux from the Proca equation also decreases exponentially, and the Coulomb potential is replaced by a Yukawa potential, [72], thus explaining the Olbers paradox and resulting in low velocity photons travelling at much less than c. The residual (combined) mass of these photons contributes to the mass of the universe and may solve the missing mass problem of cosmology [72]. The factor exp $(-\xi Z)$, such as that which appears in $B^{(3)}$ from the Proca equation (Chap. 1), implies a distance proportional red shift,

$$\frac{\Delta \nu}{\nu} \sim e^{-\xi Z}, \tag{238}$$

resulting in Tolman's "tired light". This type of red shift could contribute significantly to the cosmological red shift, and explain anomalous red shifts in objects such as quasars bound to galaxies by matter bridges. Photon mass is also consistent with anomalies in data from sources such as: double star motion, red shift discrepancies in galaxy clusters, anomalous variations in the Hubble constant, and quantized peaks in the N log z plot.

Since all these phenomena are evidently detected through telescopes, i.e., through the use of electromagnetic radiation, $B^{(3)}$ is present in all of them in one form or another, as well as in laboratory scale optical experiments of many different kinds. The field $B^{(3)}$ is a manifestation of finite photon mass, as we have argued, and is therefore fundamental in nature.

8.3 THE PROCA EQUATION

The Proca equation can be written in the form (9b) of Chap. 1, which is equivalent to

$$\frac{\partial F^{\mu\nu}}{\partial x_\mu} = -\xi^2 A^\nu \qquad (239)$$

in free space. The mass term on the right hand side is well known to be equivalent to an effective current, a term which can be derived from the free space Lagrangian term

$$\mathcal{L}_3 = -\frac{1}{2}\xi^2 A_\mu^* A^\mu, \qquad (240)$$

where ξ is given by

$$\xi = \frac{m_0 c}{\hbar}. \qquad (241)$$

The complete Lagrangian corresponding to this form of the Proca equation is therefore

$$\mathcal{L} = -\frac{\epsilon_0}{4} F_{\mu\nu}^* F^{\mu\nu} - \frac{1}{2}\xi^2 A_\mu^* A^\mu \qquad (242)$$

in S.I. units, where c and $\hbar$ are not in reduced units, as usual, and where ϵ_0 is the free space permittivity. The asterisk in this equation denotes "complex conjugate". Written out in three dimensional notation, the free space Proca equations correspond with the following, in S.I. units,

$$\nabla \cdot \mathbf{E} = -\xi^2 \phi, \quad \nabla \times \mathbf{E} = -\frac{\partial \mathbf{B}}{\partial t},$$

$$\nabla \cdot \mathbf{B} = 0, \quad \nabla \times \mathbf{B} = \frac{1}{c^2}\frac{\partial \mathbf{E}}{\partial t} - \xi^2 \mathbf{A}, \qquad (243)$$

the only difference between these and the free space Maxwell equations is that there appears a mass term in two of them, made up of the vector and scalar components of the four-vector A_μ of Eq. (236). Since m_0 is of the order 10^{-51} kgm at most [130], the difference between the free space Maxwell and Proca equations is very small for laboratory purposes (which is why the mass of the photon is so difficult to measure), but physically, they represent entirely different concepts of light in vacuo. The new $\mathbf{B}^{(3)}$ field is a phase free, constant, spin field in the Maxwell equations, but in the Proca equations, is a very slowly decaying exponential, which has a Z dependence. This is typical of the Yukawa potential in nuclear physics [72], and over cosmic dimensions, the

The Proca Equation

exponential decay in $\mathbf{B}^{(3)}$ leads to observable effects as reviewed very briefly in Sec. 8.1. If the photon is a particle with mass, m_0, it follows that its properties become different in different Lorentz frames, and that it has a rest frame. The Proca equations (243) in free space, unlike the Maxwell equations, are therefore subject to the Lorentz transformation, and so are all physical solutions thereof. To explain this important point it is worth describing briefly the properties of the free space Maxwell equations, and to show why they are Lorentz invariant.

8.3.1 LORENTZ INVARIANCE OF THE MAXWELL EQUATIONS IN FREE SPACE

In Minkowski notation, $x_\mu := (X, Y, Z, ict)$, the Maxwell equations in free space can be written as

$$\frac{\partial F_{\mu\nu}}{\partial x_\nu} = 0, \qquad (244)$$

which means that the four-divergence of $F_{\mu\nu}$ vanishes, a four-divergence which is by definition a four-vector in K. Equation (244) means that this four-vector vanishes in K. In frame K', following Jackson [4], Eq. (244) becomes

$$\frac{\partial F'_{\lambda\sigma}}{\partial x'_\sigma} = 0, \qquad (245)$$

and the transformed four-vector also vanishes in frame K'. This transformed four-vector can be given the symbol α'_λ in frame K'. By definition

$$\alpha'_\mu = a_{\mu\nu}\alpha_\nu \qquad (246)$$

under the Lorentz transformation $a_{\mu\nu}$ [4]. Therefore

$$a_{\lambda\mu}\frac{\partial F_{\mu\nu}}{\partial x_\nu} = 0 \qquad (247)$$

in frame K. We have simply back-transformed Eq. (245) (frame K') to Eq. (247) (frame K). Equation (247) must be the same as Eq. (244), and so

$$a_{\lambda\mu}\frac{\partial F_{\mu\nu}}{\partial x_\nu} = \frac{\partial F_{\mu\nu}}{\partial x_\nu} = \frac{\partial F'_{\lambda\sigma}}{\partial x'_\sigma} = 0, \tag{248}$$

showing that Maxwell's equations in frame K are the same as in frame K' in free space.

8.3.2 COVARIANCE OF THE PROCA EQUATION IN FREE SPACE

Applying the same procedure to the Proca equation in the form (239), it is seen that under the Lorentz transformation

$$a_{\lambda\mu}\frac{\partial F_{\mu\nu}}{\partial x_\nu} = -\xi^2 a_{\lambda\mu} A_\mu \tag{249}$$

and the equations become different in different Lorentz frames. Applying the Lorentz transformation to individual electric and magnetic fields, and to the four-vector A_μ in Eq. (249), it is seen that the right and left hand sides will transform as

$$\frac{\partial F'_{\lambda\sigma}}{\partial x'_\sigma} = -\xi^2 A'_\lambda \tag{250}$$

from frame to frame, and the equation will therefore depend on the velocity v of the frame K' with respect to the frame K.

8.4 ANALOGY BETWEEN PHOTON MASS AND EFFECTIVE CURRENT

We now write the Proca equation in the form

$$\Box A_\mu = \xi^2 A_\mu = -J_\mu^{(eff)}, \tag{251}$$

where $J_\mu^{(eff)}$ is an effective current four-vector. This corresponds to Eqs. (243) if we use the Minkowski notation,

Analogy Between Photon Mass and Effective Current

$$J_\mu^{(eff)} := \left(\frac{\mathbf{J}^{(eff)}}{c}, i\rho^{(eff)}\right), \qquad A_\mu := (\epsilon_0 c\mathbf{A}, i\epsilon_0 \phi), \qquad (252)$$

so that there is an effective current three-vector and charge defined by

$$\mathbf{J}^{(eff)} = -\xi^2 c^2 \epsilon_0 \mathbf{A}, \qquad \rho^{(eff)} = -\xi^2 \epsilon_0 \phi. \qquad (253)$$

These equations show that the photon mass is equivalent to the presence in free space of an effective charge and current. The product

$$A_\mu J_\mu^{(eff)} = -\xi^2 A_\mu A_\mu \qquad (254)$$

is an invariant free space electromagnetic energy density, and with our new gauge condition

$$A_\mu A_\mu = 0 \qquad (255)$$

introduced in Eq. (145) Chap. 4 we arrive at the important result that *photon mass adds nothing to the free space electromagnetic energy density*. The condition (255) is equivalent to assuming that the four-vector A_μ is physically meaningful, and has longitudinal and time-like components as in any four-vector. From the cyclic relations (25b) of Chap. 1 we see that these components are pure imaginary, whereas the transverse components are in general complex waves, as usual, with real and imaginary parts. These inferences are consistent with the Aharonov-Bohm effect [15] which shows experimentally that the four-vector A_μ is physically meaningful.

The condition $A_\mu A_\mu = 0$ has the important consequence that finite photon mass is made consistent with gauge invariance of unified field theory.

The Proca equation can be written in covariant contravariant notation as [54]

$$F^{\mu\nu} = \partial^\mu A^\nu - \partial^\nu A^\mu, \qquad \partial_\mu F^{\mu\nu} + \xi^2 A^\nu = 0, \qquad (256)$$

and the inhomogeneous Maxwell equations as [54]

$$F^{\mu\nu} = \partial^\mu A^\nu - \partial^\nu A^\mu, \qquad \partial_\mu F^{\mu\nu} - j^\nu = 0, \qquad (257)$$

so that it becomes clear that the term $-\xi^2 A^\nu$ in the Proca equation *in free space* is analogous with the current term in the inhomogeneous Maxwell equations *in matter*. In both the Maxwell and Proca equations the four-tensor $F^{\mu\nu}$ is the four-curl of the four-vector A_μ. In three dimensional notation

$$\mathbf{B} = \nabla \times \mathbf{A}, \qquad \mathbf{E} = -\nabla \phi - \frac{\partial \mathbf{A}}{\partial t}. \tag{258}$$

A simple solution to Eq. (243d) can be obtained if it is assumed that $\partial \mathbf{E}/\partial t = 0$, i.e., that the electric field has no time dependence, whereupon

$$\nabla \times \mathbf{B} = -\xi^2 \mathbf{A}, \tag{259}$$

which becomes the equation

$$\nabla \times (\nabla \times \mathbf{A}) = -\nabla^2 \mathbf{A} = -\xi^2 \mathbf{A}, \tag{260}$$

using the relation (258) between $\mathbf{A}$ and $\mathbf{B}$. Using the classical quantum equivalence

$$\nabla := \frac{i}{\hbar} \mathbf{p}, \qquad \frac{\partial}{\partial t} := -\frac{i}{\hbar} En, \tag{261}$$

results in the equation

$$En^2 - c^2 p^2 = c^2 \hbar^2 \xi^2, \tag{262}$$

which is Einstein's equation of motion of special relativity with

$$\xi = \frac{m_0 c}{\hbar}. \tag{263}$$

The relation (253a) between effective current and vector potential for finite m_0 is analogous with the London equation of superconductivity [52],

$$\mathbf{J} = -k^2 \mathbf{A}, \tag{264}$$

which is the theoretical basis for the well known Meissner effect in which magnetic flux density ($\mathbf{B}$) is excluded from the interior of a superconductor. The equation (260) leads with Eq. (258) to the relation

$$\nabla^2 \mathbf{B} = \xi^2 \mathbf{B}, \tag{265}$$

a physical solution of which is

$$\mathbf{B} = \mathbf{B}^{(0)} e^{-\xi Z}, \tag{266}$$

showing that the longitudinal magnetic field $\mathbf{B}^{(3)}$ from the Proca equation is damped exponentially. In precise analogy, the longitudinal $\mathbf{B}$ in the Meissner effect is damped exponentially in a skin depth k^{-1}, i.e., [52]

$$\mathbf{B} = B^{(0)} \exp(-kZ). \tag{267}$$

The damping in the Meissner effect takes place very quickly, but the damping due to ξ in Eq. (266) occurs very slowly, because ξ is a minute quantity.

Mathematically, the Proca equations in free space have the same form of solution as the Maxwell equations in the presence of charges and currents, photon mass produces an effective free space current four-vector $J_\mu^{(eff)}$.

8.5 GENERAL SOLUTIONS OF THE PROCA EQUATION

Using the fact that finite photon mass introduces an effective current $J_\mu^{(eff)}$ it is possible to find general solutions of the Proca equation using the standard methods devised for the inhomogeneous Maxwell equations in the form of the d'Alembert equation. The four-current of the d'Alembert equation is replaced by the effective current of the Proca equation. Methods for solving the d'Alembert equation in matter are well developed and are described in a contemporary textbook such as that of Jackson [4]. For example the description of the Liénard-Wiechert equations given by Jackson in his Chap. 14 can be adopted directly by assuming that the effective current $J_\mu^{(eff)}$ caused by photon mass can be further described in terms of a localized effective charge and current distribution without boundary surfaces. Converting to S.I. units and using Jackson's notation and the relation (251) between A_μ and the effective four-current, the general solution for A_μ from the Proca equation is

$$A_\mu = -\frac{\epsilon_0 \xi^2}{4\pi} \int\int \frac{A_\mu}{R} \delta\left(t' + \frac{R(t')}{c} - t\right) dt', \qquad (268)$$

where $\boldsymbol{R} = (\boldsymbol{Z} - \boldsymbol{r}(t)')$ and the delta function provides the usual retarded behavior given the Liénard-Wiechert potentials. The effective current of the Proca equation is given in this model by

$$J_\mu^{(eff)} = -\xi^2 A_\mu, \qquad (269)$$

in S.I. units, where ϵ_0 is the permitivity of free space. If the finite photon mass is thought of as producing an effective point charge $e^{(eff)}$ moving with velocity $\boldsymbol{\beta}$ at the point $r(t)$ the charge-current density is

$$J_\nu^{(eff)} = e^{(eff)} c \beta_\mu \delta(\boldsymbol{Z} - \boldsymbol{r}(t)), \qquad (270)$$

where $\beta_\mu = (\boldsymbol{\beta}, i)$. This line of reasoning introduces a fundamental conceptual link between the mass of the photon and the effective elementary charge $e^{(eff)}$ in free space. Within this concept of effective elementary charge $e^{(eff)}$, the four-potential from the Proca equation can be written as

$$A_\mu = e^{(eff)} \int \frac{\beta_\mu(t')}{R(t')} \delta\left(t' + \frac{R(t')}{c} - t\right) dt', \qquad (271)$$

which is directly analogous with Jackson's equation (14.3), which uses the elementary charge on the electron, e, in preference to the effective elementary charge $e^{(eff)}$ produced by photon mass. Mathematically, however, Eqs. (14.3) of Jackson and Eq. (271) are structurally identical.

Treating the Proca equation in this way has the double advantage of allowing a direct identification of photon mass with effective charge and of allowing solutions of the Proca equation using the well developed techniques [4] of electrodynamics in matter, described by the d'Alembert equation with "real" four-current J_μ. It is important to note however that the discovery of the cyclically symmetric equations (4) means that solutions of the Proca equation for magnetic fields in free space must produce $\boldsymbol{B}^{(3)}$ as well as $\boldsymbol{B}^{(1)}$ and $\boldsymbol{B}^{(2)}$. It has been shown already in this chapter that the equation does indeed produce a longitudinal $B^{(3)}$ with a very slow exponential decay in free space. Conventionally in field theory the Proca equation is described as a massive vector field [15] equation, which for the massive photon becomes a massive spin-one field with three independent components. The Lorentz condition implied by the Proca equation is used to "eliminate" one of the components of A_μ, namely the time-like component. This procedure replaces the full covariant term $A_\mu A_\mu$ in the Lagrangian by the space-like $-\mathbf{A} \cdot \mathbf{A}$, and so is equivalent to assuming arbitrarily that $A^{(0)}$, the time-like component of A_μ, is zero. This appears to be an unjustified assumption in field theory, and the existence of the cyclic relations (4) bears this out. It is clear that from Eqs. (25b) of Chap. 1, the cyclic equations for components of A_μ, that the time-like part is pure imaginary, and *not* zero. As explained in that chapter, the cyclic relations between components of $\mathbf{A}$ are self-dual to the cyclic relations among components of $\mathbf{B}$. In the previous chapter we have discussed, furthermore, the existence of experimental evidence for the effect of $B^{(3)}$, and it follows that there is experimental evidence which contradicts the arbitrary assumption that the time-like component of A_μ is zero. There is therefore a need to re-examine the approach of conventional field theory to the Proca equation, which involves $\hbar$ through the mass parameter ξ, and is therefore an equation of the quantized field theory.

We have argued that the four-vector A_μ is physically meaningful, and that gauge invariance of the second kind is satisfied by the light-like condition

$$A_\mu A_\mu = 0. \qquad (272)$$

If this is so, then A_μ must be physically meaningful, and so it is important to argue that there is experimental evidence that supports this view. That this is indeed the case is explained in the next chapter, devoted to the Aharonov-Bohm effect and to the expected optical Aharonov-Bohm effect due to the novel and physical $B^{(3)}$ in free space.

CHAPTER 9. AHARONOV-BOHM EFFECTS

In previous chapters it has been argued that the four-vector A_μ should be a fully (or manifestly) covariant four-vector in the theory of special relativity, and therefore a physically meaningful quantity. Experimental evidence for this assertion has been available for over thirty years, in the Aharonov-Bohm effect. Essentially, this is the shift in an electron diffraction pattern caused by the space-like part of A_μ, the vector potential $\mathbf{A}$, in which terms the magnetic flux density is described by $\mathbf{B} = \nabla \times \mathbf{A}$. If the space-like part of A_μ is physically meaningful, then we assert that in general, so is the complete four-vector A_μ. The cyclic relations (25) show that the longitudinal and time-like parts of A_μ are pure imaginary, and in general the transverse parts are complex. From the light-like relation $A_\mu A_\mu = 0$, it follows that the magnitude of the space-like $\mathbf{A}$ must be equal to that of the time-like $A^{(0)}$. Therefore the time-like $A^{(0)}$ cannot be zero as asserted in the conventional approach to field theory.

It is important to provide a description in this chapter of the Aharonov-Bohm effect because the observed electron diffraction shift provides evidence for the physical reality of A_μ, and therefore for the physical reality of the cyclic relations (25b) which tie together the transverse and longitudinal components of A_μ. Aharonov and Bohm [131], in introducing the effect theoretically, argued that contrary to the conclusions of classical electrodynamics, A_μ affects the trajectories of charged particles even in regions where the magnetic and electric fields are excluded. The prediction was shortly afterwards verified experimentally by Chambers [132], using the shift in the fringe pattern in an electron interference experiment. The Chambers experiment has subsequently been repeated independently several times, and the effect has become useful in mapping flux lines in contemporary superconductor technology. It is proven beyond reasonable doubt, therefore, that the space-like $\mathbf{A}$ is physically meaningful.

Chapter 9. Aharonov-Bohm Effects

9.1 THE ORIGINAL THEORY OF AHARONOV AND BOHM

In this section we follow the original paper by Aharonov and Bohm [131] in order to explain the original intention of the theory. The basis of the argument is that in the quantum theory the canonical formalism is necessary and the potential functions $\mathbf{A}$ and $A^{(0)}$, the space-like and time-like parts of A_μ, cannot be excluded as in the classical theory of fields. Despite the fact that the equations of quantum field theory are gauge invariant, Aharonov and Bohm [130] argued that the potentials $\mathbf{A}$ and $A^{(0)}$ have physical significance. These authors calculated the phase difference due to the interference of two electron beams, and expressed it as the integral around a closed circuit in space-time,

$$(S_1 - S_2) = e \oint \left(\phi dt - \frac{\mathbf{A}}{c} \cdot d\mathbf{x} \right). \tag{273}$$

It is therefore clear that the integral considered by Aharonov and Bohm involved from the outset both the time-like and space-like components of the four-vector A_μ, which was considered manifestly covariant and physically meaningful. The original theory was then specialized by Aharonov and Bohm to the space-like case: "As another special case, let us now consider a path in space only (t = constant)."

In contemporary descriptions such as that due to Ryder [54] the effect is attributed to the space-like $\mathbf{A}$, and described in terms of the interference pattern set up by two electron beams in a double slit experiment. A solenoid is placed between the two slits, a solenoid contained within which is a magnetic field $\mathbf{B}$. The latter is therefore prevented from interacting with the two electron beams, made up of electrons conventionally described by a wave function,

$$\psi = |\psi| \exp\left(\frac{i}{\hbar} \mathbf{p} \cdot \mathbf{r}\right) := |\psi| \exp(i\alpha), \tag{274}$$

where $\mathbf{p}$ is the linear momentum of one electron and where $\mathbf{r}$ is its position.

The interference pattern is then changed by an amount which depends on the vector potential $\mathbf{A}$ due to the field $\mathbf{B}$, even though the latter is excluded from the electron beams, the nature of the interaction being described essentially by the requirement of gauge invariance which leads to

$$\alpha \to \alpha - \frac{e}{\hbar}\mathbf{A}\cdot\mathbf{r}. \tag{275}$$

The phase of the wave changes according to

$$\Delta\alpha = -\frac{e}{\hbar}\int \mathbf{A}\cdot d\mathbf{r}, \tag{276}$$

and the change in phase over an entire trajectory is

$$\Delta\alpha_1 = -\frac{e}{\hbar}\int_1 \mathbf{A}\cdot d\mathbf{r}, \quad \Delta\alpha_2 = \frac{e}{\hbar}\int_2 \mathbf{A}\cdot d\mathbf{r}. \tag{277}$$

The change in the phase difference between the trajectories of the two electron beams is then calculated to be,

$$\Delta\delta = \Delta\alpha_1 - \Delta\alpha_2 = \frac{e}{\hbar}\Phi, \tag{278}$$

where Φ is the flux through the solenoid. The interference pattern of the two electron beams therefore moves upwards by an amount,

$$\Delta x = \frac{L\lambda}{2\pi}\Delta\delta = \frac{L\lambda}{2\pi d}\frac{e}{\hbar}\Phi. \tag{279}$$

The solenoid therefore causes a shift in the interference pattern even though the field $\mathbf{B}$ is excluded from the region of electron beam interference.

In a fully relativistic treatment of the same effect, the space-like part $\mathbf{A}$ is replaced by the four-potential A_μ, as in the original paper by Aharonov and Bohm, and as in subsequent treatments, notably that by Wu and Yang [132]. It is clear therefore that the original and subsequent theories of the Aharonov-Bohm effect use a manifestly covariant four-potential A_μ, i.e., an A_μ in which it is implicitly assumed that all four components are physically meaningful. This approach seems to contradict the assumptions field theory made, for example, in the quantization scheme conventionally adapted for the Proca equation, a scheme in which the time-like component of A_μ is conventionally set to zero. These difficulties reverberate throughout electromagnetic field theory, because it is conventionally assumed that A_μ has only two (transverse) components. Following Ryder [54], for example: "The origin of these difficulties is that the electromagnetic field, like any massless field, possesses only two independent components, but is covariantly described by a 4-vector A_μ. In choosing two of these components as the physical ones, and thence quantizing them, we lose manifest covariance. Alternatively, if we wish to keep covariance, we have two redundant components."

9.2 THE EFFECT OF THE CYCLIC ALGEBRA (25)

It is clear that the novel cyclic algebra (25) means that the electromagnetic field has more than just two independent components. For example there are three space-like components $\boldsymbol{B}^{(1)}$, $\boldsymbol{B}^{(2)}$ and $\boldsymbol{B}^{(3)}$, three space-like components $\boldsymbol{E}^{(1)}$, $\boldsymbol{E}^{(2)}$ and $i\boldsymbol{E}^{(3)}$, and three space-like components $\boldsymbol{A}^{(1)}$, $\boldsymbol{A}^{(2)}$ and $i\boldsymbol{A}^{(3)}$. The field component $\boldsymbol{B}^{(3)}$ is real and physically meaningful, and the component $i\boldsymbol{E}^{(3)}$ is imaginary and not physically significant. For finite photon mass to be compatible with gauge invariance of the second kind, the condition $A_\mu A_\mu = 0$ must be used, a condition which means that the time-like part of A_μ is non-zero. We have also argued that there exist time-like components of the free space electric and magnetic fields of electromagnetism, components which can be related to the electromagnetic four-tensor $F_{\mu\nu}$ in free space. The longitudinal and time-like components of the electric field are related as in Eq. (25c) and are both pure imaginary. The equivalents for the magnetic field are both pure real. These findings mean that there are more than two independent components of A_μ, and that manifest covariance is not lost in the theory of electromagnetism. However, the new cyclic relations (25) also mean that a novel, rigorously self-consistent, potential formalism for electromagnetism in free space must be constructed, one which allows for the possibility of longitudinal as well as transverse components of fields in free space. Before attempting this, however, we describe the optical Aharonov-Bohm effect due to $\boldsymbol{B}^{(3)}$, the spin field of electromagnetism in free space.

9.3 THE OPTICAL AHARONOV-BOHM EFFECT

The fundamental idea behind the optical Aharonov-Bohm effect (OAB) is the simple one of replacing the magnetic flux density of the solenoid by the field $\boldsymbol{B}^{(3)}$ of a circularly polarized laser beam. It is easy to show that only a few watts of laser power passed through a micron optical fiber produces a measurable fringe shift, using $\boldsymbol{B}^{(3)}$ in Eq. (279)

Optical Aharonov-Bohm Effect

calculated with $|\boldsymbol{B}^{(3)}| \sim 10^{-7} I_0^{\frac{1}{2}}$. The OAB, if carried out, will show whether $\boldsymbol{B}^{(3)}$ acts at first order, as in ONMR, described in Chap. 7. If so the fringe shift should be proportional to the square root of laser intensity and change sign with the sense of circular polarization of the laser. There should also be an OAB effect in elliptical polarization but none in linear polarization or if the light beam is incoherent. If observed and developed there might also be several uses to which the OAB could be put, because the rather delicate procedure of setting up an iron whisker, or small solenoid, between the two slits of the electron diffraction apparatus [131] would be replaced by a laser beam passed through a micron radius optical fiber.

The OAB is a much more sensitive test for $\boldsymbol{B}^{(3)}$ at first order than the three techniques briefly overviewed in Chap. 7. The OAB simultaneously and accurately tests for the ability of $\boldsymbol{B}^{(3)}$ to act at first order, in addition to its second order effects as evidenced in magnetization effects of light (Chap. 7). It is therefore a radically new type of experiment in electrodynamics, both classical and quantum. Essentially, a circularly polarized laser beam is expected to produce a Δx according to

$$\Delta x = \frac{L\lambda}{d} \frac{e}{\hbar} \int \boldsymbol{B}^{(3)} \cdot d\boldsymbol{s}, \tag{280}$$

where λ is the wavelength of the electron beam entering the two slits, L is the distance between the screen containing the slits and detector, d is the inter-slit separation and $\int \boldsymbol{B}^{(3)} \cdot d\boldsymbol{s}$ is a surface integral. If an optical fiber is used the beam is confined to it and the free space value of $\boldsymbol{B}^{(3)}$ is given by

$$|\boldsymbol{B}^{(3)}| \sim 10^{-7} I_0^{\frac{1}{2}}. \tag{281}$$

From these equations, estimates show that a laser beam of half a mm radius and about a milliwatt in power should provide an easily observable Δx.

9.4 THE PHYSICAL A_μ AND FINITE PHOTON MASS

Not only would an observed OAB demonstrate the ability of $\boldsymbol{B}^{(3)}$ to act at first order, but as argued in previous chapters it would also be an indirect but firm indication that the photon as particle has three polarizations in free space, two transverse and one longitudinal. Such a property is indicative of finite photon mass, i.e., of a massive spin one vector field. The experimental evidence for finite photon mass was reviewed in Chap. 8, and has important consequences for contemporary gauge theory. The observation of the effect of $\boldsymbol{B}^{(3)}$ at first order would be strongly indicative of the fact that the elementary particle known as the photon has mass. The observation of an OAB effect would, more precisely, indicate the existence of a vector potential function defined through its relation to $\boldsymbol{B}^{(3)}$,

$$\boldsymbol{B}^{(3)} = \nabla \times \boldsymbol{A}_3 . \tag{282}$$

This vector potential function would be part of the complete four-vector A_μ of free space electromagnetism. In the Einstein-de Broglie theory of light, this wave vector has physical significance, in that it pilots, or controls, the motions and trajectories of particle-like photons which carry the energy $E = h\nu$. The photons in this view follow real space-time paths tangential to the field's conserved four-vector density, associated with Madelung fluid elements [72]. The Einstein-de Broglie theory of light therefore makes interesting analogies with the hydrodynamic theory of fluids. Both Einstein and de Broglie assumed that all directly observable effects and interactions of the A_μ field result from the impact and emission of point like particles which are identified as photons. The continuous A_μ field is only indirectly observable through its effects on its associated particle like photons. In an analogous manner, the A_μ field in the AB and OAB effects displaces the electron interference pattern in a two slit experiment. Inherent in the Einstein-de Broglie theory of light is the acceptance of finite (i.e., rigorously non-zero, or positive definite) photon mass. As first shown by Bass and Schrödinger [133] and by Moles and Vigier [134], finite photon mass leads to the modifications in free space electromagnetism discussed in Chap. 8, one of the most important of which is the fact that the Proca equation naturally allows longitudinal solutions in free space. As first shown by de Broglie [135] these longitudinal solutions correspond to a Yukawa potential which approximates the Coulomb field. In this volume we have argued for the existence of a *novel* longitudinal $\boldsymbol{B}^{(3)}$ field in free space, a field which is phase free, and which is a very slowly decaying exponential solution of the Proca equation. It is important to note that the original solutions by de Broglie, Bass and Schrödinger, and Moles and Vigier, corresponded to a phase dependent, electric, longitudinal fields, and that Moles and Vigier in their original paper set $\boldsymbol{B}^{(3)}$ =? 0 arbitrarily. However, their solution of the Proca equation also allows a non-zero $\boldsymbol{B}^{(3)}$ as discussed already.

In Chap. 1 it was shown that the existence in free space of $\boldsymbol{B}^{(3)}$ and $-i\boldsymbol{E}^{(3)}/c$, its dual in special relativity, makes no difference to the electromagnetic energy density in free space. Products such as $F^{\mu\nu}F^*_{\mu\nu}$, (Eq. (7)) remain zero. However, the presence of a photon mass term in the Lagrangian produces an extra electromagnetic energy density as shown in Eq. (254) of Chap. 8. It is important to understand therefore that $\boldsymbol{B}^{(3)}$ exists for zero photon mass as well as for finite photon mass, but that the third (longitudinal) degree of polarization indicated by the presence of $\boldsymbol{B}^{(3)}$ in free space does not result in an increase by a factor 3/2 in the Planck constant. The extra contribution to electromagnetic energy density of a mass term in the Lagrangian is minute, which is one of the reasons why the Proca and d'Alembert equations produce practical results which are virtually indistinguishable.

9.5 IF THE OAB IS NOT OBSERVED

If the OAB is not observed experimentally it would mean that $\boldsymbol{B}^{(3)}$ does not act at first order, and therefore does not generate a vector potential in free space. This enigmatic result would mean that the magnetization described in Eq. (6) must always be produced by a product of $\boldsymbol{B}^{(3)}$ with its own amplitude $B^{(0)}$. Therefore $\boldsymbol{B}^{(3)}$ would be a quantity which has the units and fundamental symmetries of magnetic flux density, but which would not act at first order as such. Failure to observe $\boldsymbol{B}^{(3)}$ in the OAB would mean that another mechanism would have to be sought to explain the optical NMR results of Warren et al. [113, 116], which, as described in Chap. 8, are enormously greater in magnitude than allowed for in conventional, second order, perturbation theory. The mechanism put forward by Warren et al. themselves [113] time averages to zero and seems not to explain the observed shifts. As discussed in Chap. 8, the only first order magnetic field of free space electromagnetism that does not time average to zero is $\boldsymbol{B}^{(3)}$. The fields $\boldsymbol{B}^{(1)}$ and $\boldsymbol{B}^{(2)}$, as used by Warren et al. [113] are phase dependent and at first order average to zero. At the optical frequencies used in the ONMR experiment, no shift would be observed due to $B^{(1)}$ and $B^{(2)}$ at first order, and shifts at second order appear to be many orders of magnitude too small.

Non-observation of an OAB would appear therefore to be fundamentally incompatible with the results of the ONMR experiment unless the latter has produced artifact. In Chap. 7 it was argued that this is unlikely.

Making, nevertheless, the unlikely assumption that the series of ONMR experiments at Princeton has produced artifact, and assuming that an OAB is not observable, we would be driven in the light of the reasoning that has led to equations (4) and (6) to the conclusion that free space electrodynamics, based on the d'Alembert or Proca equations, is not self-consistent, in that the cyclic Lie algebra (4) produces a self-contradiction or paradox. If $B^{(1)}$ and $B^{(2)}$ are physical fields, then the Lie algebra (4) shows that $B^{(3)}$ must also be a physical field in classical or quantum electrodynamics in free space. Therefore if $B^{(3)}$ is not observed experimentally as a magnetic field, acting as such at first order, electrodynamics has contradicted itself at the most fundamental level. At the time of writing there appears to be no satisfactory alternative explanation, some attempted criticisms have been reviewed in Chap. 7. As mentioned there, these seem to us to be subjective in nature, i.e., do not lead to a satisfactory objective explanation of why $B^{(1)}$ and $B^{(2)}$ should be physical, and why $B^{(3)}$ should be unphysical, if indeed, experiments show it to be so. At the time of writing there are very few incisive data available, despite the enormity of the contemporary literature, so the question of $B^{(3)}$ is open and the photon remains an enigma.

Referring to Eq. (6), describing in simplified terms the inverse Faraday effect, it appears beyond reasonable experimental doubt that the IFE: a) exists; b) has been observed to date to be proportional to the product $B^{(0)}B^{(3)}$. This shows that if $B^{(3)}$ were not a magnetic field, the quantity $M^{(3)}$ would not be a magnetization and would therefore not be observable experimentally as a magnetization. The contrary occurs however, and so by this reasoning, $B^{(3)}$ is an *observed* magnetic field. Similar conclusions hold from the optical Faraday effect, as discussed in Chap. 7. Fundamentally, it is unsurprising that the various observed magnetic effects of light should have been produced by a magnetic field, and it is equally clear that the original explanation for the inverse Faraday effect, given in terms of $B^{(1)} \times B^{(2)}$,

If the OAB is Not Observed

can be very simply re-expressed as $iB^{(0)}\boldsymbol{B}^{(3)}$. It follows that arguments such as those reviewed in Chap. 7 that assert that $\boldsymbol{B}^{(3)}$ is zero lead immediately to the disappearance of $\boldsymbol{B}^{(1)} \times \boldsymbol{B}^{(2)}$ and therefore to the disappearance of the IFE.

In previous chapters, arguments have also been given for the fundamentally geometrical nature of $\hat{B}^{(3)}$, which has been shown to be directly proportional to the rotation generator $\hat{J}^{(3)}$, an obviously physical quantity, which generates a physical rotation in three dimensions. Since $\hat{B}^{(3)}$ is simply $\hat{J}^{(3)}$ multiplied by $iB^{(0)}$, (Eq. (107)), it is directly self-contradictory to assert that $\hat{B}^{(3)}$ is anything other than a physically meaningful magnetic flux density. The OAB appears to be a relatively straightforward test of this assertion. The algebra (4) is a Lie algebra because its components are directly proportional to physical rotation generators, which form a well-defined group in space, and by extension of this reasoning, in space-time. In the quantum theory $\hat{B}^{(3)}$ becomes the photomagneton operator which is directly proportional to the longitudinal component in free space of the photon angular momentum, another obviously physical quantity, which as obviously acts at first order upon matter. (Photon angular momentum is transferred in well defined integral multiples of $\hbar$ when light interacts with matter, either by ordinary absorption or in magnetization far from resonance, the inverse Faraday effect.) It has also been argued that $\boldsymbol{B}^{(3)}$ does not affect the fundamental theory of light, for example $\boldsymbol{B}^{(3)}$ and its dual, as shown in Chap. 1, add nothing to free space electromagnetic energy density, and therefore Planck's law is unchanged. Again, $\boldsymbol{B}^{(3)}$ in free space is a solution of the d'Alembert equation, and therefore of the four Maxwell equations in free space, equations which define the physical meaning of a field. The structure of the well known electromagnetic four-tensor $F_{\mu\nu}$ allows for the existence of a longitudinal $\boldsymbol{B}^{(3)}$ and its imaginary dual: $\boldsymbol{B}^{(3)}$ is relativistically invariant (i.e., is not affected by the Lorentz transformation applied to $F_{\mu\nu}$) and is fully compatible with special relativity in free space.

Since photon angular momentum is well known to act at first order on matter, it is also expected that a physically meaningful $\hat{B}^{(3)}$ would act likewise. This supposition can be investigated in more detail by writing

$$\hat{B}^{(3)}(0; -\omega, \omega) = \frac{B^{(0)}}{\hbar} \hat{J}(0; -\omega, \omega). \tag{283}$$

The field $B^{(3)}$ in its classical interpretation depends for its existence on the cross product $B^{(1)} \times B^{(2)}$ of negative and positive frequency transverse modes (1) and (2) which are complex conjugate pairs. This is the fundamental geometrical reason why the basis (1), (2) and (3) can be used as an alternative to the Cartesian basis (X, Y, Z). It follows that the photon angular momentum itself must be generated from the same phase free product of negative and positive frequency waves. However, it is clear that this angular momentum acts on matter at first order, and so must $\hat{B}^{(3)}$, the physical magnetic field, on the basis of this reasoning. That $\hat{B}^{(3)}$ is indeed capable of doing so is supported by data from ONMR [113, 116].

Using all these arguments therefore, it is expected that there should be an OAB due to $\hat{B}^{(3)}$, and that it would be paradoxical if none were observed.

9.6 NON-LINEARITY OF PHOTON SPIN IN FREE SPACE

The eigenvalues of the massless photon are well known to be $\hbar$ or $-\hbar$, but the classical definition of angular momentum of light, given in Eq. (44) of Chap. 1, is a volume integral over the product $E^{(1)} \times B^{(2)}$ and is therefore non-linear, in that it is a product of fields with positive and negative phase coefficients. Similar reasoning leads to the conclusion that the field $B^{(3)}$ is also non-linear in nature, because, as we have argued in earlier chapters, it is always formed from the experimentally observable cross product $B^{(1)} \times B^{(2)}$. Similarly, the Stokes operators of the quantum field theory are formed from bilinear products of electric fields but are also angular momentum operators [15]. However, the magnetic fields $B^{(1)}$, $B^{(2)}$ and $B^{(3)}$ are angular momentum operators which are to first order in the magnetic field, and as in previous argument, it is unreasonable to assert that $B^{(3)}$ is unphysical if $B^{(1)}$ and $B^{(2)}$ are taken to be physical, in the same way that it is unreasonable to assert that one Stokes operator out of four is unphysical. The field $B^{(3)}$ is directly proportional to S_3, and so if $B^{(3)}$ were unphysical, so would S_3, in direct contradiction with standard theory [4, 15].

However, it remains true that the interaction of $B^{(3)}$ with matter must reflect its fundamental character, i.e., account for the fact that it is defined (Chap. 3) as:

Non-Linearity of Photon Spin in Free Space 143

$$\hat{B}^{(3)} := \hat{B}^{(3)}(0; -\omega, \omega). \tag{284}$$

Similarly,

$$\hat{J} := \hat{J}(0; -\omega, \omega), \quad \hat{S}_3 := \hat{S}_3(0; -\omega, \omega). \tag{285}$$

Any description of $\hat{B}^{(3)}$ as "static" must therefore reflect the fact that it has no net (i.e., functional) dependence on phase, $\phi = \omega t - \kappa \cdot r$. In the same way, $\hat{J}$ and $\hat{S}_3$ have none. For a given beam intensity in circular polarization, S_3 is a constant of magnitude $\pm E^{(0)2}$, + for left and - for right circular polarization.

The Stokes operators and magnetic field operators are both angular momentum commutators in free space, and can both be described in terms of bilinear products of creation and annihilation operators (Chap. 5 and 6). In the Copenhagen interpretation, the three field components cannot be specified simultaneously, as usual in angular momentum theory in quantum mechanics. This is consistent with the fact that the (3) (or Z) component of photon angular momentum is usually specified in eigenvalue form, eigenvalues which are longitudinal projections $\hbar$ and $-\hbar$. For the photon with mass, the eigenvalues are $\hbar$, 0 and $-\hbar$, the usual eigenvalues of the spin-one boson. In special relativity, furthermore, the transverse angular momentum components for a massless particle travelling at c are mathematically indeterminate, and there is no rest frame. In contrast, the longitudinal component is relativistically invariant. Thus, in classical special relativity, angular momenta components behave under Lorentz transformation as

$$J_Z = J_Z', \quad J_Y = \gamma J_Y', \quad J_X = \gamma J_X', \tag{286}$$

where $\gamma := (1 - v_Z^2/c^2)^{-\frac{1}{2}}$. If the relative velocity of two frames is c, then J_X and J_Y (in the static, observer frame) become infinite unless $J_X' = J_Y' = 0$, in which condition J_X and J_Y are indeterminate mathematically but in which $J_Z = J_Z'$ is well defined and invariant. The same overall result is obtained in the quantum theory, the field operators $\hat{B}^{(1)}$ and $\hat{B}^{(2)}$ are not specified if $\hat{B}^{(3)}$ is specified.

Therefore $\hat{B}^{(3)}$ is also invariant, and is an angular momentum operator which is intrinsically non-linear in nature. Nevertheless, if it is a magnetic field as conventionally accepted, it must act as such on matter. Some signs of this are apparent in the prototype ONMR experiment, but the results are not yet unequivocal because there is a complicated dependence on the light intensity, the cause of which is not known (Chap. 7).

If we compare directly the classical and quantum equations

$$\mathbf{B}^{(1)} \times \mathbf{B}^{(2)} = iB^{(0)}\mathbf{B}^{(3)}, \qquad \text{and cyclics,}$$

$$[\hat{B}^{(1)}, \hat{B}^{(2)}] = -iB^{(0)}\hat{B}^{(3)}, \qquad \text{and cyclics,} \qquad (287)$$

it becomes immediately obvious that Eq. (287) is a relation between spins in the Maxwellian interpretation. Each spin component (1), (2) and (3) is formed from a vector cross product of the other two; this being a requirement of Euclidean geometry. In order for this geometrical requirement to satisfy simultaneously Maxwell's equations in free space, in particular the equation $\nabla \cdot \mathbf{B}^{(3)} = 0$, the longitudinal component $\mathbf{B}^{(3)}$ must be phase free, otherwise its divergence is non-zero because the phase has a Z dependence. In order to satisfy this and the other three Maxwell equations, the transverse components $\mathbf{B}^{(1)}$ and $\mathbf{B}^{(2)}$ must be phase dependent. Equations (4) tie these considerations together in a circular basis, in the same way that rotation generators in classical theory and angular momenta in quantum theory are tied together.

The field $\hat{B}^{(3)}$ is also a constant of motion, being a phase free angular momentum operator in free space, while $\hat{B}^{(1)}$ and $\hat{B}^{(2)}$ are governed by photon statistics and are subject to purely quantum effects such as light squeezing, as discussed in previous chapters. The field $\hat{B}^{(3)}$, as we have seen, is not subject to light squeezing and its eigenvalues remain constant in free space. This result is consistent with the fact that in quantum mechanics the general expression for the rate of change of an expectation value is

$$\frac{d}{dt}\langle\hat{B}^{(3)}\rangle = \frac{i}{\hbar}\langle[\hat{H}, \hat{B}^{(3)}]\rangle, \qquad (288)$$

and $\hat{B}^{(3)}$ commutes with the Hamiltonian operator $\hat{H}$, as in angular momentum theory. This is consistent with the fact that $\mathbf{B}^{(3)}$ and its imaginary dual add nothing to light intensity, and are frequency independent in nature, so have

Non-Linearity of Photon Spin in Free Space

no light quantum energy $h\nu$ in any conventional meaning. In other words the expectation value of $\hat{B}^{(3)}$ although formed from the non-linear product $\mathbf{B}^{(1)} \times \mathbf{B}^{(2)}$, is independent of time, and its eigenvalues are constant. Similarly, the Stokes operator $\hat{S}_3$ to which $\hat{B}^{(3)}$ is directly proportional is also a constant of motion and independent of electromagnetic phase and time. The spin of the massless photon is $\pm\hbar$, and the photomagneton $\hat{B}^{(3)}$ is a direct consequence of photon spin. The classical $\mathbf{B}^{(3)}$ is therefore a direct consequence of the fact that there exists right and left circular polarization in electromagnetic radiation.

As in Chap. 3, the Heisenberg uncertainty principle shows that

$$\delta\hat{B}^{(1)}\delta\hat{B}^{(2)} \geq \frac{1}{2}|B^{(0)}\hat{B}^{(3)}| \qquad (289)$$

where $\delta\hat{B}^{(1)}$ and $\delta\hat{B}^{(2)}$ are root mean square deviations. The product on the right hand side is a rigorous lower bound in the quantum theory, a lower bound on the nonlinear product $\delta\hat{B}^{(1)}\delta\hat{B}^{(2)}$, a lower bound which is defined in terms of $\hat{B}^{(3)}$. Thus, if $\hat{B}^{(3)}$ were zero, $\hat{B}^{(1)}$ and $\hat{B}^{(2)}$ would commute in the quantum theory, implying that $\delta\hat{B}^{(1)} = 0$ and $\delta\hat{B}^{(2)} = 0$ simultaneously. The experimental observation of light squeezing shows that the quantum theory is valid in electromagnetic radiation, and produces effects which are not describable in the classical field theory, effects such as squeezing and anti-bunching [15]. On this basis, therefore, light squeezing is a firm experimental indication of the existence of the product of $B^{(0)}$ and $\hat{B}^{(3)}$, as is Eq. (6). Since $B^{(0)}\hat{B}^{(3)}$ vanishes if $\hat{B}^{(3)}$ is zero, it appears clear that $\hat{B}^{(3)}$ is finite, and that arguments to the contrary are incorrect. If $\hat{B}^{(3)}$ is finite and rigorously non-zero, and has the units and discrete symmetries ($\hat{C}$, $\hat{P}$, and $\hat{T}$) of a magnetic flux density, it is such in field theory.

If the result of the proposed OAB experiment is negative despite these arguments, it will have to be asserted that $\hat{B}^{(3)}$ does not generate a vector potential in free space which is capable of affecting electron interference patterns. The overall conclusion of such a negative result will be a paradox in electrodynamics at a fundamental level, a paradox which would have to be addressed by a modification of the Maxwell equations themselves. These equations are empirically based, and as such are built on experimental observation.

If they are contradicted by novel experimental data, they must be modified in response, and this is of course an interesting development.

If a positive result to the OAB experiment is observed experimentally, it would imply that $\hat{B}^{(3)}$ is a magnetic field that generates a vector potential in free space. The net result would be to strengthen electrodynamics as currently understood in terms of the Maxwell equations in free space, and their modifications for finite photon mass discussed in Chap. 8. It appears at the time of writing therefore that the OAB effect is a key experiment for electrodynamics, and as interesting and challenging as the original AB effect. The latter overturned the long-accepted notion that A_μ is mathematically convenient but unphysical, and has led to many novel insights as described, for example, by Ryder [52, 54].

CHAPTER 10. MODIFICATIONS OF LAGRANGIAN FIELD THEORY

The novel cyclic algebra (25) requires modifications to the conventional Lagrangian theory of fields, because the electric part of free space electromagnetism is described in that theory in terms of a four-vector,

$$\pi_\mu = \frac{1}{i}\epsilon_0 E_\mu, \qquad (290)$$

and is identified with the conjugate momentum π_μ. In Eq. (290) we have used Minkowski's notation and the i is not suppressed as in covariant-contravariant notation. The Lagrangian field theory is used conventionally [54] in the Lorentz gauge quantization of the electromagnetic field in free space, and the conjugate momentum is defined through the Lagrange equation,

$$\pi_\mu = \frac{\partial \mathscr{L}_E}{\partial(\partial A_\mu/\partial x^{(0)})}, \qquad x^{(0)} := ict, \qquad (291)$$

where $\mathscr{L}_E$ is the field Lagrangian and $\partial A_\mu/\partial x^{(0)}$, the time derivative of the four-potential. Comparing Eq. (291) with the simple equivalent Lagrange equation of classical dynamics in Cartesian coordinates,

$$P_j = \frac{\partial \mathscr{L}}{\partial \dot{q}_j}, \qquad (292)$$

it is seen that A_μ is a generalized position and π_μ a generalized conjugate momentum. Using Eq. (290) it is seen that in the conventional Lagrangian field theory of electromagnetism there exists the concept of an electric field four-vector E_μ which is the generalized momentum conjugate to the generalized "position" A_μ. Thus, if A_μ is manifestly covariant, with four components in free space, then so is E_μ. The cyclically symmetric equations (25) of Chap. 1 show that both A_μ and E_μ have pure imaginary longitudinal components in free space, the magnitude of which contributes to the time-like component. The scalar magnitude of an imaginary quantity in the theory of complex numbers is a real quantity, so the light-like condition $A_\mu A_\mu = 0$, obtained from the requirement of gauge invariance with finite photon mass, means that A_μ has a pure imaginary, phase free, longitudinal component, and a real, non-zero, time-like component equal to the real magnitude of its space-like part.

The conventional theory asserts, however, that the longitudinal space-like components of A_μ and E_μ vanish, and removes the time-like components from consideration. Following the description, for example, of Lorentz gauge quantization by Ryder [54] the conjugate momentum fields in his contravariant-covariant notation are given by

$$\pi^\mu = \frac{\partial \mathcal{L}}{\partial \dot{A}_\mu}, \qquad \pi^0 = 0, \qquad (293)$$

and although π^μ is in general a four-vector of special relativity, its time-like component is evidently discarded from the outset as being zero, and the space-like component is identified with a purely space-like electric field. This leads to the conventional view that the electromagnetic field in free space has two transverse components, unrelated to an assumed unphysical longitudinal component in free space.

Equations (4) and (25) render this point of view untenable, because the transverse components are linked geometrically to the longitudinal, pure real, $\boldsymbol{B}^{(3)}$, and the longitudinal, pure imaginary, $i\boldsymbol{E}^{(3)}$, as argued in detail already, $-i\boldsymbol{E}^{(3)}/c$ being the dual of $\boldsymbol{B}^{(3)}$ in special relativity. This duality property means that if $\boldsymbol{B}^{(3)}$ is non-zero and real, then there exists a non-zero, imaginary longitudinal electric field in free space, a field which is a solution of Maxwell's equations.

The conventional Lagrangian field theory also runs immediately into trouble when quantization of the electromagnetic field is attempted in the Lorentz gauge, because the conventional Lagrangian (in covariant-contravariant notation),

$$\mathcal{L} = -\frac{1}{4} F_{\mu\nu} F^{\mu\nu}, \qquad (294)$$

leads to the result, in the same notation [54],

$$\pi^0 = \frac{\partial \mathcal{L}}{\partial \dot{A}_0} = 0 \qquad (295)$$

and A_0, the time-like part of A_μ, commutes with π^0, the time-like part of π^μ. This means that A_0 becomes in this view a c-number, following Ryder [54], and not an operator, and covariance is lost, meaning that the theory is not compatible with the second principle of special relativity. This is remedied conventionally by changing the Lagrangian with the well known gauge fixing technique, well described by Ryder [54]. For our present purposes, it is worth emphasizing that the gauge fixing term must be introduced in order to provide a non-zero π^0, i.e., a non-zero time-like component of π^μ and therefore of E^μ in free space. This is precisely what is indicated by Eqs. (25) and the condition $A_\mu A_\mu = 0$ (in Minkowski notation), introduced on the assumption of finite photon mass. In the conventional theory [54] however, the Feynman gauge fixing term simply leads back to

$$\pi^0 = -\partial_\mu A^\mu = 0, \qquad (296)$$

π^0 again vanishes, and the theory again becomes incompatible with special relativity in that it loses covariance. This problem is addressed conventionally with the Gupta-Bleuler condition, which is less rigorous than the Lorentz condition, and which leads in turn to the deduction that physical states are admixtures of longitudinal and time-like photon states as discussed in earlier chapters. The Gupta-Bleuler approximation of the Lorentz condition does not therefore allow the existence of independent longitudinal and time-like photon states, and this conflicts with the cyclically symmetric conditions (4) and (25), which are geometrical in nature. There is therefore a need for a fundamental re-appraisal of the conventional approach to field quantization in view of Eqs. (4) and (25), an approach which takes in its stride the existence of non-zero photon mass. The conventional approach to quantization of the massive spin-one field, as described by the Proca equation, yet again leads back to the result $\pi^{(0)} = 0$, i.e., to the disappearance of the time-like component of the conjugate momentum π_μ and thus of the electric field four-vector E_μ. Therefore, the inclusion of a photon mass term in the Lagrangian does not in itself lead to the desired result indicated by the condition $A_\mu A_\mu = 0$, i.e., that the magnitude of the space-like part of A_μ is equal to its time-like part. Any Lagrangian theory that leads to the deduction that the time-like part of A_μ is zero means that *its space-like part is also zero*, and in view of Eqs. (4) and (25), this is untenable experimentally and theoretically. If the time-like and space-like parts of A_μ vanish, there is no electromagnetism in vacuo.

10.1 NOVEL GAUGE FIXING TERM

Using Minkowski's notation, the Lagrangian in Eq. (294) is modified to

$$\mathscr{L}_E = -\frac{1}{4} F_{\mu\nu} F_{\mu\nu} - \frac{\partial A^{(0)}}{\partial x^{(0)}} \frac{\partial A_\mu}{\partial x_\mu}, \qquad A^{(0)} := \frac{i\phi}{c}, \qquad (297)$$

for the photon without mass, i.e., modified with a novel gauge fixing term which replaces the usual $-1/2 (\partial A_\mu/\partial x_\mu)^2$. Using the Lorentz condition

$$\frac{\partial A_\mu}{\partial x_\mu} = 0, \qquad (298)$$

it is seen that the second term in Eq. (297) adds nothing to the Lagrangian, and it has the required dimensions, symmetry, and scalar character. It is also covariant, i.e., is consistent with special relativity, and is invariant to gauge transformations of the second kind because it is always zero by the Lorentz condition. It replaces the Feynman gauge fixing term because it has the necessary property of leading to an independent time-like component of E_μ in free space,

$$\epsilon_0 E^{(0)} = \pi^{(0)} = -\frac{\partial A^{(0)}}{\partial x^{(0)}}, \qquad (299)$$

where ϵ_0 is the free space permittivity. This time-like component is thereby identified as $E^{(0)}$, and this also appears in the fully covariant four-tensor $F_{\mu\nu}$ as discussed in Chap. 3. As defined by Eq. (299), the time-like component is real, an inference which is consistent with the fact that it is the magnitude of the space part of the four-vector E_μ in free space, i.e.,

$$E_\mu E_\mu = 0, \qquad \boldsymbol{E} \cdot \boldsymbol{E} - E^{(0)2} = 0. \qquad (300)$$

We arrive at the important result that the time-like component of E_μ is the scalar magnitude of the electric component of free space electromagnetism. This inference is

Novel Gauge Fixing Term

consistent with the fact that the conjugate momentum π_μ is properly a four-vector in space-time, i.e., a manifestly covariant, physically meaningful concept. The novel Lagrangian (297) restores meaning to π_μ, which becomes

$$i\pi_\mu := (\pi_i, i\pi^{(0)}), \tag{301}$$

and is fully covariant and fully consistent with special relativity. Equation (297) is consistent with the d'Alembert equation of motion, because the latter is equivalent to the Lagrange equation of motion,

$$\frac{\partial \mathcal{L}_E}{\partial A_\mu} = \frac{\partial}{\partial x_\nu} \left(\frac{\partial \mathcal{L}_E}{\partial (\partial A_\mu / \partial x_\nu)} \right). \tag{302}$$

We have

$$\frac{\partial \mathcal{L}_E}{\partial (\partial A_\mu / \partial x_\nu)} = -\left(\frac{\partial A_\mu}{\partial x_\nu} - \frac{\partial A_\nu}{\partial x_\mu} \right) - g_{\mu\nu} \frac{\partial A^{(0)}}{\partial x^{(0)}}, \quad \frac{\partial \mathcal{L}_E}{\partial A_\mu} = 0, \tag{303}$$

where $g_{\mu\nu}$ is the Minkowski metric tensor, which vanishes for $\mu \neq \nu$. For $\mu = \nu = 0$, then $g_{00} = 1$, and

$$\pi^{(0)} = \frac{\partial \mathcal{L}_E}{\partial (\partial A^{(0)} / \partial x^{(0)})} = -\frac{\partial A^{(0)}}{\partial x^{(0)}} = \epsilon_0 E^{(0)}, \tag{304}$$

which is Eq. (299). The space-like part of π_μ (and therefore of E_μ) is given from Eq. (303) by setting $\nu = 0$, $\mu \neq 0$, so $g_{\mu 0} = 0$, and

$$-i\pi_i = \frac{\partial \mathcal{L}_E}{\partial (\partial A_i / \partial x^{(0)})} = \frac{\partial A^{(0)}}{\partial x_i} - \frac{\partial A_i}{\partial x^{(0)}} = -i\epsilon_0 E_i, \tag{305}$$

$$\pi_i = \epsilon_0 E_i.$$

There is therefore a simple proportionality between a manifestly covariant four-vector π_μ and E_μ, proving that the latter is conjugate to A_μ in Lagrangian dynamics.

This result has been arrived at through an appropriate choice of gauge fixing term in the Lagrangian of the Euler-Lagrange equation of motion of the classical electromagnetic field in vacuo. The gauge fixing term adds zero to the original Lagrangian because we are working within the Lorentz gauge, and using the Lorentz condition (298). This link between the fully covariant A_μ and the fully covariant E_μ is a result of the equation of motion given the new gauge fixing term, and the method shows that E_μ contains four components in free space. The scalar potential $A^{(0)} = i\phi/c$ (in S.I.) cannot be set to zero as in the conventional theory.

The Feynman gauge fixing term is chosen conventionally so that $\pi^{(0)} = 0$ is a result, meaning that the time-like $E^{(0)}$ has no independent existence, an inference that results from the Gupta-Bleuler method of field quantization in the Lorentz gauge.

The d'Alembert equation (or, for the photon with mass, the Proca equation) is recovered from Eqs. (302) and (303) using

$$\frac{\partial}{\partial x_\nu}\left(\frac{\partial \mathcal{L}_E}{\partial(\partial A_\mu/\partial x_\nu)}\right) = -\frac{\partial}{\partial x_\nu}\left(\frac{\partial A_\mu}{\partial x_\nu} - \frac{\partial A_\nu}{\partial x_\mu}\right) - \frac{\partial}{\partial x_\nu}\left(g_{\mu\nu}\frac{\partial A^{(0)}}{\partial x^{(0)}}\right) = 0, \quad (306)$$

i.e.,

$$\frac{\partial}{\partial x_\nu}\left(\frac{\partial A_\mu}{\partial x_\nu}\right) = \Box A_\mu = 0, \quad (307)$$

and so the Lagrangian (297) is consistent with these fundamental classical field equations. The existence of the four-vector E_μ does not contradict this fundamental relation.

10.2 QUANTIZATION OF THE ELECTROMAGNETIC FIELD

Quantization of the Maxwellian field in the Lorentz gauge becomes a self-consistent procedure with Eqs. (301), (304) and (305). It is remarkable that although the enigmatic photon as light quantum, $h\nu$, is by now a concept familiar to the layman, the quantization of the Maxwellian field has been beset with difficulty. The emergence of Eqs. (4) and (25) however, over ninety years after Planck's original proposal of 1900, leads to a self-consistent Lagrangian interpretation as we have argued already. If the real field $\boldsymbol{B}^{(3)}$ is experimentally observed with more accuracy, and if the current experimental uncertainties (e.g. Chap. 7) are removed, and the lack of data remedied, it may well be that field quantization would have been made entirely self consistent and consistent with special relativity.

Quantization of the Electromagnetic Field

The position momentum equal time commutator in the quantized electromagnetic field [54] is written as

$$[\hat{A}_\mu(\mathbf{x}, t), \hat{\pi}_\nu(\mathbf{x}', t)] = ig_{\mu\nu}\delta^3(\mathbf{x} - \mathbf{x}') \tag{308}$$

in the usual way, but now, π_ν is fully covariant, having been identified through Eq. (290) with E_μ. The basic field commutator is therefore the fully covariant

$$[\hat{A}_\mu(\mathbf{x}, t), \hat{E}_\nu(\mathbf{x}', t)] = -\frac{g_{\mu\nu}}{\epsilon_0}\delta^3(\mathbf{x} - \mathbf{x}'). \tag{309}$$

It is also clear that

$$[\hat{A}_\mu(\mathbf{x}, t), \hat{A}_\nu(\mathbf{x}', t)] = [\hat{E}_\mu(\mathbf{x}, t), \hat{E}_\nu(\mathbf{x}', t)] = 0. \tag{310}$$

This commutator must be carefully distinguished from one such as Eq. (181), which commutes $\hat{E}^{(1)}$, defined as a boost generator, with $\hat{E}^{(2)}$, the complex conjugate of this boost generator. In Eq. (310), $\hat{E}_\mu$ and $\hat{E}_\nu$ are not complex conjugates.

There is a key difference, therefore, between the method proposed in this section and the traditional relativistic quantization of the electromagnetic field, in which $\pi^{(0)}$ (and thus $E^{(0)}$) have no independent physical existence. In the traditional point of view [54] the existence of $E^{(0)}$ is not recognized fully, but the space-like E_i is at the same time identified with the space like π_i. It is recognized, traditionally, that there is a four-vector π_μ conjugate to a four-vector A_μ in Lagrangian field theory, and that the space-like part of π_μ is directly proportional to the space-like E_μ, but illogically, the time-like $\pi^{(0)}$ is set to zero, and manifest covariance destroyed. This procedure is rescued with the Gupta-Bleuler method, whose key result is the inference that only admixtures of longitudinal and time-like photons can be meaningful physically, the components separately can have no independent physical meaning. In our opinion, Eqs. (4) and (25), which link together the transverse and longitudinal components of free space electromagnetism, require the abandonment of the traditional approach in favor of a more self consistent one. The simple reason for this is that the novel cyclically symmetric relations (4) and (25) show the existence of longitudinal components of the space-like E, A, and B. The longitudinal component $B^{(3)}$, having the known properties of a real magnetic field, is *physically* meaningful, thus contradicting the main result of the Gupta-Bleuler method — that only admixtures of longitudinal and time-like components can be physically meaningful.

Clearly, the time-like component of E_μ, i.e., $E^{(0)}$, should properly have the same units as its space-like component, and therefore should be proportional to electric field strength amplitude in volt m^{-1}. We have argued that π_μ must be proportional to E_μ, whose time-like part, $E^{(0)}$, is non-zero and proportional to $\pi^{(0)}$ (Eq. (304)). This leads to a basic commutator, Eq. (309), which is rigorously equivalent with the d'Alembert and Euler-Lagrange equations of the classical field. The four-vector E_μ becomes the canonical momentum of A_μ, both being rigorously covariant in vacuo. By defining E_μ as being proportional to the conjugate momentum π_μ, it is clear that E_μ must behave under Lorentz transformation in the same way as π_μ, which in turn is defined through A_μ by the Euler-Lagrange field equations. The Lagrangian of this equation contains the four-tensor $F_{\mu\nu}$, thus establishing a link between E_μ, A_μ, and $F_{\mu\nu}$.

The use of E_μ has the advantage of retaining the Lorentz equation as a meaningful operator identity, because $\partial A_\mu/\partial x_\mu$ is no longer equal to $\pi^{(0)}$ as in the traditional method [54]. If E_μ is recognized as a four-vector therefore the Lorentz condition no longer conflicts with the basic commutator relations (308) and (310) of the quantized field, and quantization becomes a self-consistent procedure. It therefore becomes debatable whether there is a further need for the traditional Gupta-Bleuler method of field quantization in the Lorentz gauge, but nevertheless, at this point in our development the essence of that method is recounted briefly, following Ryder [54].

The conventional argument [54] for the Gupta-Bleuler method of quantization of the electromagnetic field rests on the need to circumvent a particular difficulty. Therefore if this difficulty is taken away by Eqs. (4) and (25), as argued already, there is no need for the method at all. The difficulty is that the Lorentz condition $\partial A_\mu/\partial x_\mu = 0$ cannot be regarded in the traditional approach as an operator identity because it conflicts with the commutators (308). However, this conflict arises because of the use of the Lagrangian (294). With the gauge fixing term (299) introduced in this chapter, there is no longer a conflict between the Lorentz condition and the basic field commutators. This inference is a powerful result of the novel Eqs. (4) and (25) as we have

Quantization of the Electromagnetic Field

argued. The method introduced by Gupta and Bleuler follows from the conventional need to replace the Lorentz operator condition by

$$\frac{\partial \hat{A}_\mu}{\partial x_\mu} |\psi\rangle = 0, \tag{311}$$

where $|\psi\rangle$ is a physical eigenstate of the field. In terms of the traditional positive and negative frequency decomposition [54] Eq. (311) becomes

$$\frac{\partial \hat{A}_\mu^{(+)}}{\partial x_\mu} |\psi\rangle + \frac{\partial \hat{A}_\mu^{(-)}}{\partial x_\mu} |\psi\rangle = 0. \tag{312}$$

This leads to the difficulty [54] that the negative frequency operator contains creation operators, so the identity (312) as it stands is unphysical. It cannot be satisfied in a vacuum state [54]. This difficulty is met with the assertion first made in the early days of quantum theory by Gupta and Bleuler [137], an assertion which through use has become accepted uncritically,

$$\frac{\partial \hat{A}_\mu^{(+)}}{\partial x_\mu} |\psi\rangle = 0. \tag{313}$$

This equation is satisfied by the vacuum state. The inferences of Gupta and Bleuler rest on this assertion, which appears in a critical light to be made a posteriori, with the purpose of coming to the result described below that only transverse field components can be physically meaningful. The traditional theory [54] develops through logical consequences of Eq. (313), the most important of which is

$$\hat{a}^{(0)}(k)|\psi\rangle = \hat{a}^{(3)}(k)|\psi\rangle. \tag{314}$$

This is the admixture condition discussed already. From it

$$\langle\psi|\hat{a}^{(0)+}(k)\hat{a}^{(0)}(k)|\psi\rangle = \langle\psi|\hat{a}^{(3)+}(k)\hat{a}^{(3)}(k)|\psi\rangle \tag{315}$$

and the contributions of longitudinal and time-like photons to the Hamiltonian cancel. From this, it is inferred traditionally that only the transverse states contribute to the Hamiltonian, and so only they can be physically meaningful.

By reference to equations such as (7) of Chap. 1, however, it can be shown that the fields $B^{(3)}$ and $-iE^{(3)}/c$ are rigorously dual and, being parallel in all Lorentz frames, can never contribute to the light intensity (time averaged Poynting vector in vacuo). The Lorentz invariants L_1 and L_2 of the electromagnetism in vacuo remain rigorously zero in the presence of $B^{(3)}$ and its dual $-iE^{(3)}/c$. This is a classical relativistic result to which a correct quantization scheme must reduce through classical quantum equivalence. Although $B^{(3)}$ and its dual contribute nothing to L_1 and L_2 in free space, it is clear that $B^{(3)}$ has, at the same time, all the properties of a classical magnetic field, as argued in previous chapters. It has therefore a physical existence of its own, and is also longitudinal. This shows that the admixture equation (314) cannot be interpreted as meaning that longitudinal field components have no independent physical existence. In our interpretation, the longitudinal component $B^{(3)}$ is *independently* physical, and so is the timelike $B^{(0)}$. The latter is a scalar magnitude defined through the four-tensor $F_{\mu\nu}$ as discussed already. The dual field in vacuo, $-iE^{(3)}/c$, is not physical, because it is imaginary, but is nevertheless an independent field component. Its magnitude is the real and physical $E^{(0)}/c$, which is the timelike component of the free space four-vector E_μ. Similar conclusions hold for the imaginary longitudinal component of A_μ in free space and its time-like component.

To resolve these interpretations experimentally needs considerations such as those given in Chap. 7. At the time of writing, it is beyond reasonable doubt that the product $iB^{(0)}B^{(3)}$ produces magnetization in matter in the IFE, and there are signs from ONMR (Chap. 7) that $B^{(3)}$ can act at first order, although the data are far from being unequivocal. Similarly, the experimentally observed optical Faraday effect (Chap. 7) can be interpreted qualitatively in terms of $B^{(3)}$ acting to shift the MCD spectrum upwards and downwards on the frequency scale without distortion of the overall spectrum, i.e., acting as a magnetic field. It has been suggested in Chap. 9 that the optical equivalent of the well known AB effect be pursued experimentally as a sensitive test for $B^{(3)}$, and that if the OAB is negative, the Maxwell equations themselves need to be reformulated to account for the fundamental algebra (4), a Lie algebra, and at the same time to disallow the physical existence of one component, $B^{(3)}$,

Quantization of the Electromagnetic Field

even though it is part of the physical Lie algebra.

To resolve the interpretations theoretically requires objective consideration of the Lie algebra represented by Eqs. (4) and (25), an algebra which is rigorously isomorphic with that of the physical, infinitesimal boost and rotation generators of the Lorentz group in vacuo. Obviously, theory prior to the discovery of Eqs. (4) and (25) is based on the almost universal assertion that only transverse components of the electromagnetic field in vacuo can be physical fields. As obviously, the traditional approach includes the various assertions of the method of Gupta and Bleuler, assertions which appear in a critical light to have been made in order to obtain the desired result that only transverse components can be physically meaningful. In this sense therefore, the Gupta-Bleuler argument is circular.

Conventionally, it leads to the definition of the total energy of an ensemble of photons in terms of a field Hamiltonian which is an integral over the sum

$$\hat{H} \propto \sum_{\lambda=1}^{3} (\hat{a}^{(\lambda)+}\hat{a}^{(\lambda)} - \hat{a}^{(0)+}\hat{a}^{(0)}), \tag{316}$$

consisting of bilinear products (Chaps. 5 and 6) of creation and annihilation operators. The index λ over the three space-like components and the sum also includes time-like creation and annihilation operators indexed (0). The method asserts, however, that only admixtures of time-like and longitudinal photons can be physically meaningful, so that the Hamiltonian, although it contains these components explicitly, reduces to a sum over just two, transverse components, labelled (1) and (2). Ryder, for example, describes this procedure as meaning that the contributions of the longitudinal and time-like components cancel. This type of Hamiltonian arises, however, from a solution of the d'Alembert equation using a Fourier integral procedure, i.e., essentially from an integration of that equation for A_μ in terms of its Fourier components. If A_μ is to be regarded as manifestly covariant and physically meaningful, then all four of its components must contribute to the field Hamiltonian. If the magnitude of the time-like part of A_μ happens to be equal to the magnitude of its longitudinal part, then the procedure outlined by Ryder [54] remains valid. The interpretation of the theory is however fundamentally different, because both the longitudinal and time-like parts of A_μ are independently physically meaningful, and not just as admixtures as in the conventional interpretation of Gupta and

Bleuler.

To illustrate this point, we develop in the next section a potential model for A_μ which takes into account that there is a physically meaningful $B^{(3)}$ from Eqs. (4), a field $B^{(3)}$ which must be derivable from a suitable space-like vector potential using the relation $B = \nabla \times A$.

10.3 A POTENTIAL MODEL FOR $B^{(3)}$

The transverse part of the electromagnetic field can be derived from Eqs. (10) as usual. However, if there exists a real, physical $B^{(3)}$ in free space, the corresponding vector potential must also be transverse (because $B^{(3)}$ is longitudinal, and is derived from the curl of the corresponding vector potential), and must contribute to Eqs. (10). The vector potential components

$$A^{(1)} = i\frac{B^{(0)}}{\kappa} e^{(1)} e^{i\phi} + \frac{\sqrt{2}}{4} B^{(0)} e^{(1)}(iX - Y),$$

$$A^{(2)} = -i\frac{B^{(0)}}{\kappa} e^{(2)} e^{-i\phi} + \frac{\sqrt{2}}{4} B^{(0)} e^{(2)}(-iX - Y),$$

(317)

provide the fields

$$\nabla \times (A^{(1)} + A^{(2)}) = B^{(1)} + B^{(2)} + B^{(3)},$$

$$-\frac{\partial}{\partial t}(A^{(1)} + A^{(2)}) = E^{(1)} + E^{(2)},$$

(318)

in free space, and therefore give the required result that there are three physical magnetic field components and two physical electric field components. (The real part of the longitudinal electric field $E^{(3)}$ is zero.)

Equations (25), however, show that the product of $A^{(1)}$ and $A^{(2)}$ in free space gives an imaginary $iA^{(3)*}$, whose symmetry is taken to be that of an *axial* vector. Therefore the cross product of $A^{(1)}$ and $A^{(2)}$ must also be taken into consideration in examining the complete consequences of our potential model. There are two types of non-zero cross term. The first is given by

A Potential Model for $B^{(3)}$

$$iA^{(3)*} = i\frac{B^{(0)2}}{8A^{(0)}}R^2 e^{(3)*}, \quad R^2 = X^2 + Y^2 = \text{constant} \quad (319)$$

and gives an $iA^{(3)*}$ whose curl and time derivative are both zero. Therefore it contributes nothing to the electric or magnetic components of electromagnetism in free space. The second cross term gives an $iA^{(3)*}$ of the following type,

$$iA^{(3)*} = -\frac{\sqrt{2}B^{(0)2}}{4\kappa A^{(0)}}(e^{-i\phi}(iX-Y) - e^{i\phi}(-iX-Y))e^{(3)*}, \quad (320)$$

the curl of which is

$$\nabla \times (iA^{(3)*}) = \frac{1}{2c}(E^{(2)} - E^{(1)}). \quad (321)$$

The potential $iA^{(3)*}$ therefore produces from its curl a combination of *electric* fields in free space, and the real (physically meaningful) part of this combination vanishes. This result is consistent with the fact that $iA^{(3)*}$ is an axial vector from Eq. (25), so that its curl produces a polar vector, and is reminiscent of the Hertz potential method [4] in which an electric field is obtained from the curl of a vector potential, the "mirror image" of the usual method. This result suggests furthermore that the most symmetric potential representation of Eqs. (4) can be obtained with the Hertz potentials, suitably modified and generalized. The time derivative of the vector potential $--iA^{(3)*}$ is both longitudinal and a function of the electromagnetic phase,

$$\frac{\partial}{\partial t}(iA^{(3)*}) = f(e^{-i\phi}) + f(e^{i\phi}), \quad (322)$$

so that it does not represent an acceptable magnetic field in free space because such a field is not a solution of the Maxwell equation $\nabla \cdot B = 0$. The divergence of $iA^{(3)}$ cannot, furthermore, give an electric field, as for those of $A^{(1)}$ and $A^{(2)}$, because the symmetry of $-iA^{(3)*}$ is axial, while those of both $A^{(1)}$ and $A^{(2)}$ are polar. Finally, using the light-like condition $A_\mu A_\mu = 0$ derived in foregoing chapters, the scalar potential is obtained in the appropriate circular basis from

$$\phi = c|A| = c(A^{(1)} \cdot A^{(1)*} + A^{(2)} \cdot A^{(2)*} + A^{(3)} \cdot A^{(3)*})^{\frac{1}{2}}, \quad (323)$$

and its divergence vanishes. The scalar potential obtained in this way is non-zero, therefore, but does not contribute to the physically significant magnetic fields $\boldsymbol{B}^{(1)}$, $\boldsymbol{B}^{(2)}$ and $\boldsymbol{B}^{(3)}$, or to the physical electric fields $\boldsymbol{E}^{(1)}$ and $\boldsymbol{E}^{(2)}$.

In conclusion, the potential model (317), although not yet in the most symmetric form, produces the physical electric and magnetic components in free space of electromagnetism, i.e., accounts satisfactorily for the existence of the inter-related and cyclically symmetric Lie algebra (4) and (25).

CHAPTER 11. PSEUDO FOUR-VECTOR REPRESENTATIONS OF ELECTRIC AND MAGNETIC FIELDS

In previous chapters it has been argued that there exist in free space a real, physical, and longitudinal $\boldsymbol{B}^{(3)}$; $B_Z = |\boldsymbol{B}^{(3)}|$, and a time-like $B^{(0)}$. The equivalent, longitudinal $i\boldsymbol{E}^{(3)}$ is pure imaginary and is dual to $\boldsymbol{B}^{(3)}$ so that the Maxwell equations in free space are invariant to the transformation $\boldsymbol{B}^{(3)} \to -i\boldsymbol{E}^{(3)}/c$. These deductions were arrived at by constructing the Lie algebra (4) and extending this formally to Eqs. (25). Therefore there exist in free space three space-like components $\boldsymbol{B}_X$, $\boldsymbol{B}_Y$ and $\boldsymbol{B}_Z$, and the time-like $B^{(0)}$, and it is inferred that these form a type of four-vector,

$$B_\mu := \left(B_X, B_Y, B_Z, iB^{(0)}\right). \tag{324}$$

Similarly, there is, a priori, a four-vector

$$E_\mu := \left(E_X, E_Y, E_Z, iE^{(0)}\right). \tag{325}$$

The properties of these four-vectors under Lorentz transformation must be such that the individual components transform as the components of the well known four-tensor $F_{\mu\nu}$, i.e., as [4],

$$\begin{aligned}
B'_X &= \gamma\left(B_X + \frac{v}{c^2}E_Y\right), & E'_X &= \gamma(E_X + vB_Y), \\
B'_Y &= \gamma\left(B_Y - \frac{v}{c^2}E_X\right), & E'_Y &= \gamma(E_Y - vB_X), \\
B'_Z &= B_Z, & E'_Z &= E_Z,
\end{aligned} \tag{326}$$

where $\gamma = (1 - v^2/c^2)^{-\frac{1}{2}}$; and where v is the velocity of frame K' with respect to K as usual [4]. Therefore B_μ and E_μ are not ordinary four-vectors because the latter are defined through

Chapter 11. Pseudo Four-Vector Representations

the Lorentz transformation

$$x'_\mu = a_{\mu\nu} x_\nu, \qquad (327)$$

where $x_\mu := (X, Y, Z, ict)$, $X' = X$, $Y' = Y$, $Z' = \gamma(Z - vt)$, $t' = \gamma\left(t - \dfrac{vZ}{c^2}\right)$ and where

$$a_{\mu\nu} := \begin{bmatrix} 1 & 0 & 0 & 0 \\ 0 & 1 & 0 & 0 \\ 0 & 0 & \gamma & i\dfrac{v}{c}\gamma \\ 0 & 0 & -i\dfrac{v}{c}\gamma & \gamma \end{bmatrix} \qquad (328)$$

is the well known Lorentz transformation matrix. Therefore, B_μ and E_μ do not have the same transformation properties as x_μ.

The Lorentz transformation of a vector B_μ defined as in equation (324) is reminiscent of that of the angular momentum four-tensor, whose space components transform according to Eq. (286) of Chap. 9, in which the Z component is invariant. Since $\hat{B}^{(3)}$ is directly proportional to an angular momentum operator in free space, B_z is also invariant under Lorentz transformation, i.e., is independent of the relative velocity v of two inertial reference frames K and K'. These considerations lead to the inference that B_μ and E_μ are vectors that are in some way equivalent to the four-tensor $F_{\mu\nu}$. Analogously, in three dimensional Euclidean space, the axial vector of rank one is equivalent (Eq. (153)) to the antisymmetric tensor of rank two. Axial vectors, however, following Panofsky and Phillips [45], cannot form the spatial part of a four-vector such as x_μ, from which it follows that B_μ cannot behave in the same way as x_μ under Lorentz transformation. Therefore in this chapter we seek a method of defining B_μ in such a way that its components transform as required (Eq. (326)). This is achieved by finding a *pseudo four-vector* V_μ that is the sum of B_μ and E_μ in free space, and which is also dual to $F_{\rho\sigma}$ in the theory of tensors. Maxwell's equations in free space apply equally well to the components of $F_{\rho\sigma}$ and of V_μ, which has the properties of a Pauli-Lubansky pseudo-vector [54].

Minkowski and Lorentz Forces

11.1 RELATION BETWEEN THE MINKOWSKI AND LORENTZ FORCES

It is well known, following Barut [46], that the Minkowski force equation is a relativistic counterpart of Newton's equation of force and the product of mass and acceleration. The Minkowski equation is

$$K_\mu = \frac{\partial p_\mu}{\partial \tau} \tag{329}$$

where

$$p_\mu = m_0 c u_\mu \tag{330}$$

is the four-momentum (or momentum energy vector), and where

$$\partial \tau = \gamma \partial t \tag{331}$$

denotes proper time, a Lorentz invariant. Here m_0 denotes mass, which is also a Lorentz invariant, and u_μ the four-velocity of Minkowski's space-time. Following Panofsky and Phillips [45], the components of K_μ can be expressed in three dimensional notation as (Minkowski notation):

$$K_\mu := \gamma\left(\boldsymbol{K}, \frac{i\boldsymbol{K}\cdot\boldsymbol{u}}{c}\right), \tag{332}$$

where $\boldsymbol{K}$ is ordinary three dimensional force and where $\boldsymbol{u}$ is the ordinary three dimensional particle velocity. When $\boldsymbol{K}$ is proper in the inertia reference frame K, where $\boldsymbol{u} = \boldsymbol{0}$, then

$$K_\mu = (\boldsymbol{K}, 0). \tag{333}$$

In the K' frame moving at v in Z relative to K, $u_Z = 0$; $u'_Z = v$; and

$$K'_Z = K_Z, \quad K'_Y = \frac{K_Y}{\gamma}, \quad K'_X = \frac{K_X}{\gamma}. \tag{334}$$

The factor γ appears in the X and Y components of frame K' because the proper time is used in the Minkowski equation of motion. It is seen that the Z component of the force $\boldsymbol{K}$ is invariant if the particle velocity is defined to be zero in

Chapter 11. Pseudo Four-Vector Representations

frame K, the rest frame. Therefore, in this respect, the space component $\boldsymbol{K}$ of Minkowski's force K_μ transforms in the same way as angular momentum. The latter in special relativity has the transformation properties

$$J'_Z = J_Z, \quad J'_Y = \frac{J_Y}{\gamma}, \quad J'_X = \frac{J_X}{\gamma}. \tag{335}$$

Comparing Eqs. (334) and (335) it becomes clear that $K_\mu = (\boldsymbol{K}, 0)$ transforms from frame K to K' in the same way as the space components of angular momentum. The latter in Minkowski space is represented as an antisymmetric four-tensor, well described by Barut [46], and so in general K_μ is a representation of the antisymmetric four-tensor of space-time. Therefore K_μ is a pseudo four-vector of the Pauli-Lubansky type [54].

The link between the pseudo-vector and tensor representations of K_μ is given by the Lorentz force equation, as described by Barut [46],

$$K_\mu = \frac{e}{m_0 c^2} \cdot \frac{1}{2} \epsilon_{\mu\nu\rho\sigma} F_{\rho\sigma} p_\nu := \frac{e}{m_0 c^2} \cdot \frac{1}{2} G_{\mu\nu} p_\nu, \tag{336}$$

where e and m_0 are the electron's charge and mass, and where p_μ is the electronic four-momentum. Here $F_{\rho\sigma}$ is the electromagnetic four-tensor in free space. The Lorentz force equation in the form (336), where $\epsilon_{\mu\nu\rho\sigma}$ is the four dimensional, totally antisymmetric, unit tensor, is an example of the generally valid tensorial relation [46]

$$S_\mu = \frac{1}{2} \epsilon_{\mu\nu\sigma\rho} p_\nu \sigma_{\sigma\rho} \tag{337}$$

in Minkowski's notation. Equation (337) can be inverted, following Barut [46], to give

$$\sigma_{\mu\nu} = \frac{1}{2} \epsilon_{\mu\nu\sigma\rho} p_\sigma S_\rho \tag{338}$$

and S_μ is defined thereby as being *dual* to the tensor $\sigma_{\sigma\rho}$, a property which implies

$$p_\mu S_\mu = p_\mu \sigma_{\mu\nu} = S_\mu \sigma_{\mu\nu} = 0, \quad \sigma_{\mu\nu} = -\sigma_{\nu\mu} \tag{339}$$

in the theory of tensors [46]. A comparison of Eqs. (336)

Minkowski and Lorentz Forces

and (337) shows that K_μ is dual to $F_{\rho\sigma}$ in the theory of electromagnetism and matter.

11.2 DUAL PSEUDO FOUR-VECTOR OF $F_{\rho\sigma}$ IN FREE SPACE

The photon in free space is a massless particle for which $K_\mu = 0$ and

$$P_\nu = \hbar\kappa(0, 0, 1, i) \qquad (340)$$

in Minkowski's notation, is its four-momentum. Here κ is the scalar magnitude of the wave-vector as usual. It is therefore possible to define a unit momentum four-vector

$$\epsilon_\nu := \frac{P_\nu}{\hbar\kappa} = \frac{cP_\nu}{\hbar\omega} = (0, 0, 1, i) \qquad (341)$$

for the free photon. There is no Minkowski or Lorentz force because the free photon does not interact with an electron, and the four-momentum of the free photon is constant and light-like in free space. However, $\epsilon_{\mu\nu\rho\sigma}$ and ϵ_ν are both dimensionless and so the vector defined by

$$V_\mu := -\frac{i}{2}\epsilon_{\mu\nu\rho\sigma}F_{\rho\sigma}\epsilon_\nu \qquad (342)$$

has the same dimensions as $F_{\rho\sigma}$, the free space electromagnetic four-tensor. In other words V_μ is dual to $F_{\rho\sigma}$ and is composed of electric and magnetic field components in free space. Therefore Maxwell's equations in free space are invariant under the transformation

$$F_{\rho\sigma} \to V_\mu. \qquad (343)$$

From Eqs. (339)

$$\epsilon_\mu V_\mu = 0, \qquad F_{\mu\nu} = -F_{\nu\mu}. \qquad (344)$$

Following Ryder [54], V_μ is a Pauli-Lubansky pseudo four-vector with the light-like properties

$$V_\mu V_\mu = V_\mu \epsilon_\mu = \epsilon_\mu \epsilon_\mu = 0, \qquad (345)$$

Chapter 11. Pseudo Four-Vector Representations

so that the equation

$$\hat{V}_\mu |k\rangle \propto \lambda \hat{\epsilon}_\mu |k\rangle \qquad (346)$$

where $|k\rangle$ is an eigenstate of the electromagnetic field, defines the *helicity* of the massless photon, denoted by the dimensionless number λ. With considerations [54] of parity

$$\lambda = \pm 1. \qquad (347)$$

It is also clear that ϵ_ν as defined is the unit translation generator of the Poincaré (or inhomogeneous Lorentz) group of space-time [54]. Finally, V_μ is directly proportional to ϵ_μ.

11.3 LINK BETWEEN V_μ, B_μ AND E_μ

It now becomes straightforward to establish the nature of B_μ and E_μ in terms of V_μ. In S.I. units, using Minkowski's notation,

$$F_{\rho\sigma} = \epsilon_0 \begin{bmatrix} 0 & cB_Z & -cB_Y & -iE_X \\ -cB_Z & 0 & cB_X & -iE_Y \\ cB_Y & -cB_X & 0 & -iE_Z \\ iE_X & iE_Y & iE_Z & 0 \end{bmatrix}, \qquad (348)$$

which is dual to the tensor

$$G_{\mu\nu} = \frac{1}{2}\epsilon_{\mu\nu\rho\sigma}F_{\rho\sigma} = \epsilon_0 \begin{bmatrix} 0 & -iE_Z & iE_Y & cB_X \\ iE_Z & 0 & -iE_X & cB_Y \\ -iE_Y & iE_X & 0 & cB_Z \\ -cB_X & -cB_Y & -cB_Z & 0 \end{bmatrix} \qquad (349)$$

in free space. From Eqs. (349) in (342),

$$V_\mu := -iG_{\mu\nu}\epsilon_\nu = \epsilon_0(E_Y + cB_X, -E_X + cB_Y, cB_Z, icB_Z)$$

$$= \epsilon_0(0, 0, cB_Z, icB_Z), \qquad (350)$$

$$W_\mu := -iF_{\mu\nu}\epsilon_\nu = -i\epsilon_0(-cB_Y + E_X, cB_X + E_Y, E_Z, iE_Z)$$

$$= -i\epsilon_0(0, 0, E_Z, iE_Z),$$

Link Between V_μ, B_μ and E_μ 167

in free space. Therefore we arrive at our final result,

$$V_\mu = \epsilon_0(cB_\mu + (E_Y, -E_X, 0, 0)),$$

$$W_\mu = -i\epsilon_0(E_\mu + c(-B_Y, B_X, 0, 0)),$$
(351)

where

$$E_\mu = (E_X, E_Y, E_Z, iE_Z), \qquad B_\mu = (B_X, B_Y, B_Z, iB_Z). \tag{352}$$

The sum of two pseudo four-vectors is another pseudo four-vector so that E_μ and B_μ are *both* of this type. This result was first derived [15] in the recent literature using a different route. Note that B_z is dual to $-iE_z/c$ and if the former is real, the latter is imaginary.

11.4 SOME PROPERTIES OF E_μ AND B_μ IN FREE SPACE

(1) An important indication that $B_z (= |\boldsymbol{B}^{(3)}|)$ is non-zero in free space is given by combining Eqs. (346) and (350), using the definition of ϵ_v in Eq. (341). This procedure shows that B_z is proportional to the helicity λ of the free photon. Since $\lambda = \pm 1$ if there is no photon mass considered, then B_z is non-zero as argued in detail in previous chapters.

(2) Conversely, the conventional view that $B_z =? 0$ or is otherwise unphysical leads to the result that the photon helicity is also zero or unphysical, a clearly incorrect result which shows that the papers by Barron [53], Lakhtakia [103], and Grimes [104] come to an incorrect conclusion. The helicity of the free photon is a physical, non-zero, quantity, and so is B_z.

(3) The pseudo four-vector W_μ has no real longitudinal component, but from the Lie algebra (25) there exists, mathematically, a pure imaginary but unphysical electric field denoted iE_z. This leads to the definition of

$$E_\mu^{//} := (0, 0, iE_z, i|iE_z|), \tag{353}$$

a pure imaginary pseudo four-vector in free space. Note that $-iE_z/c$ is also proportional to the helicity of the free photon.

Chapter 11. Pseudo Four-Vector Representations

(4) Eqs. (345) hold for V_μ and ϵ_μ as defined, because V_μ is directly proportional to ϵ_μ through the scalar Lorentz invariant $\epsilon_0 c B_z$, an invariant which is proportional directly to the helicity of the free photon. Furthermore $\epsilon_\mu \epsilon_\mu$ is a Casimir invariant of type one (mass invariant of the Poincaré group [54]) and $V_\mu V_\mu$ is a Casimir invariant of type two (spin invariant of the Poincaré group). Both invariants are zero for the photon without mass. That $V_\mu V_\mu$ is a spin invariant means that V_μ has the properties of angular momentum [54]. Therefore so have E_μ and B_μ.

(5) For the free photon without mass, V_μ is light-like: it is a null vector in Minkowski space-time, and its norm, or metric, is zero. (This does not mean that all its components are separately zero.) The Lorentz transformation leaves it invariant, i.e., produces another zero norm. This means that B_z is Lorentz invariant and non-zero. Following Barut [46], a vector orthogonal to a light-like vector is space-like, unless it is proportional to the light-like vector. Since ϵ_μ as defined is also light-like for the free photon without mass, then V_μ must be proportional to ϵ_μ as argued already.

(6) If V_μ is dual to $F_{\rho\sigma}$ then, loosely speaking, it contains the same information as $F_{\rho\sigma}$. However, Eq. (350) shows that the real longitudinal B_z is present in V_μ, but there is no real longitudinal electric field component in V_μ. This is precisely as argued in previous chapters, there is a real, physical B_z but no real E_z. Considerations of discrete symmetry (Chap. 2) lead to the same conclusion, which can be checked by using the generally valid equations (337) to (339) of tensor theory. If V_μ is dual to $F_{\rho\sigma}$, as defined in Eq. (342), then Eqs. (339) imply rigorously that,

$$V_\mu \epsilon_\mu = W_\mu \epsilon_\mu = 0, \qquad (354)$$

which is true if and only if B_z, the real, longitudinal space component is non-zero identically. Since W_ν is simply proportional to ϵ_ν, we can easily check that $W_\mu \epsilon_\mu = 0$ as required.

(7) Therefore V_μ is a pseudo four-vector in the theory of tensors, being dual to the antisymmetric four-tensor $F_{\rho\sigma}$ in free space. Maxwell's equations in free space are rigorously invariant to the duality

transformation,

$$F_{\rho\sigma} \to V_\mu, \quad G_{\rho\sigma} \to W_\mu. \tag{355}$$

11.5 CONSEQUENCES FOR THE FUNDAMENTAL THEORY OF FREE SPACE ELECTROMAGNETISM

The above argument leads rigorously to a non-zero B_z, because it shows that B_z, the magnitude of $\boldsymbol{B}^{(3)}$ in free space, is directly proportional to the photon helicity λ. This is a conclusive demonstration that the conventional view, in which B_z is considered to be zero, or otherwise unphysical, irrelevant or similar, is in need of modification. The duality transformation (355) is also rigorously consistent with the cyclically symmetric Euclidean space relations (4) and (25). The most important consequence is the obvious one that B_z should produce physically measurable effects, as described in Chaps. 7 and 9. If it does not, then the Maxwell equations themselves would be in need of modification as discussed in Chap. 9. It is important to note that B_z is not a consequence of a "model", or empirical construct, but *a fundamental part of the theory of special relativity*, with its concomitant mathematical machinery. In this sense, therefore, B_z is a prediction of special relativity, a prediction as fundamental as the Fitzgerald-Lorentz contraction; time dilation; relativistic Doppler shifts; and so on. Therefore if B_z is not observed experimentally an enigmatic failure of special relativity itself would have occurred. If B_z is observed unequivocally, the theory of special relativity, which incorporates Maxwell's equations, would have been strengthened and shown to be self consistent.

11.6 CONSEQUENCES FOR THE THEORY OF FINITE PHOTON MASS

In previous chapters it has been argued that the existence of $\boldsymbol{B}^{(3)}$ and that of finite photon mass are interrelated, the existence of photon mass leads to a very slow exponential decay for $\boldsymbol{B}^{(3)}$ in free space. For all practical purposes in laboratory experiments, V_μ is the same for zero and non-zero photon mass. The key argument in this context is that the existence of the longitudinal $\boldsymbol{B}^{(3)}$ provides the photon in free space with three dimensionality, so that

Chapter 11. Pseudo Four-Vector Representations

quantization becomes straightforward, as described, for example, by Ryder [54], using the Proca equation for a spin-one field. Quantization of the two dimensional photon of conventional theory is beset with difficulty, both in the Coulomb and Lorentz gauges, and the classical relativistic theory leads to the result that the little group of the Poincaré group is the two dimensional Euclidean group E(2). The latter is meaningless in three dimensional Euclidean space. Indeed, it appears with hindsight that the notion of a two dimensional particle (the conventional free photon) in three dimensional space is anomalous. The enigma is that this notion has become a cornerstone of thought and has been accepted uncritically, despite its obviously unphysical nature. It has been argued in previous chapters that the difficulties are removed if it is realized that the existence of $B^{(1)}$ and $B^{(2)}$ implies that of $B^{(3)}$ in free space. In this chapter it has been shown, using the rigorous theory of tensors, that the conventional tensor $F_{\rho\sigma}$ is dual to the pseudo four-vector V_μ which is a sum of electric and magnetic pseudo four-vectors. In free space, furthermore, this sum reduces to

$$V_\mu = \epsilon_0 c(0, 0, B_z, iB_z), \qquad (356)$$

whose space-like part is purely longitudinal. This appears, finally, to be convincing evidence of the physically transparent nature of finite photon mass, a notion that is consistent with contemporary gauge theory with the condition $A_\mu A_\mu = 0$. It is important to note that A_μ is a four-vector, while B_μ and E_μ are pseudo four-vectors which form part of the pseudo-vector V_μ which is dual to the antisymmetric four-tensor $F_{\rho\sigma}$. The latter is the four-curl of A_σ as usual, so that V_μ and A_σ are interrelated. This is consistent with the fact that E and B are defined in terms of A and the scalar potential ϕ.

CHAPTER 12. DERIVATION OF $B^{(3)}$ FROM THE RELATIVISTIC HAMILTON-JACOBI EQUATION OF e IN A_μ

In previous chapters the various indications have been described for the existence of the novel spin field $B^{(3)}$, the expectation value of the photomagneton $\hat{B}^{(3)}$. In this chapter it is shown that the angular momentum of an electron (e) in a circularly polarized electromagnetic field (A_μ) can be derived from the fundamental relativistic Hamilton-Jacobi equation of motion of one electron in the four-potential A_μ. The electromagnetic property that governs the motion of the angular momentum of the electron, and therefore the induced magnetic dipole moment, is $B^{(3)}$. Under certain circumstances, to be defined, the induced magnetic dipole moment is dominated by the term in $B^{(3)}$ and therefore in the *square root* of the electromagnetic intensity in watts per square meter. Under other circumstances the dominant term is that in $B^{(3)2}$, and in general, all powers of $B^{(3)}$ contribute. It is emphasized that these conclusions can be arrived at directly from the equation of motion of e in A_μ, leaving no reasonable doubt as to the existence of $B^{(3)}$ in free space. Indeed, $B^{(3)}$ is the only field property involved in permanent magnetization by light, which is a phase independent phenomenon. This chapter is based on the excellent account of the relativistic Hamilton-Jacobi equation of e in A_μ given by Landau and Lifshitz [5], with the key additional inference that the equation predicts the existence of $B^{(3)}$ directly from the rotational trajectory of an electron in a circularly polarized electromagnetic field. In a frame in which there is initially no net electronic linear momentum, this trajectory is a circle. In this chapter, S.I. units and Minkowski's notation are used throughout. In Landau and Lifshitz, Gaussian units are used with contravariant, covariant notation.

12.1 ACTION AND THE HAMILTON-JACOBI EQUATION OF MOTION

Our derivation of $B^{(3)}$ in this chapter starts with the principle of least action, which is defined for a particle in the electromagnetic field by [5],

$$S = \int_a^b (-m_0 c\, ds + eA_\mu\, dx_\mu), \qquad (357)$$

where m_0 is the relativistically invariant particle mass, where the infinitesimal ds is

$$ds = d(x_\mu x_\mu)^{\frac{1}{2}}, \qquad (358)$$

and where the classical electromagnetic field is represented by the four-vector A_μ. The principle of least action states that the variation of the action S is zero,

$$\delta S = 0. \qquad (359)$$

The basis of this principle is philosophical in nature, it is an axiom of classical mechanics. Thus, the use of the principle of least action to show the existence of $B^{(3)}$ would succeed in proving its physical reality at the fundamental level in natural philosophy. If the four-vector A_μ is defined as

$$A_\mu := \left(\mathbf{A},\, i\frac{\phi}{c}\right), \qquad (360)$$

(as in Landau and Lifshitz [5]) then Eq. (357) in three dimensional notation becomes

$$S = \int_a^b (-m_0 c\, ds + e\mathbf{A}\cdot d\mathbf{r} - e\phi\, dt), \qquad (361)$$

and since

$$S = \int_{t_1}^{t_2} L\, dt, \qquad (362)$$

this procedure identifies the Lagrangian of e in A_μ as

$$L = -m_0 c^2 \left(1 - \frac{v^2}{c^2}\right)^{\frac{1}{2}} + e\mathbf{A}\cdot\mathbf{v} - e\phi. \qquad (363)$$

The generalized relativistic momentum $\mathbf{P}$ is a derivative of the Lagrangian,

Action and the Hamilton-Jacobi Equation

$$P := \frac{\partial L}{\partial v} = p + eA = mv\left(1 - \frac{v^2}{c^2}\right)^{-\frac{1}{2}} + eA, \quad (364)$$

and adds the term eA to the relativistic momentum p.

The generalized relativistic momentum P and Lagrangian are used in the definition of the Hamiltonian for the interaction of e and A_μ,

$$H = v \cdot \frac{\partial L}{\partial v} - L, \quad (365)$$

a Hamiltonian which can be written as

$$H = \left(m_0^2 c^4 + c^2 (P - eA)^2\right)^{\frac{1}{2}} + e\phi. \quad (366)$$

The relativistic Hamilton-Jacobi equation of motion of e in A_μ is obtained through the equations

$$P = \frac{\partial S}{\partial r}, \quad H = -\frac{\partial S}{\partial t}, \quad (367)$$

which relate P and H to derivatives of action S. In three dimensional notation the relativistic Hamilton-Jacobi equation of motion of e in A_μ is

$$(\nabla S - eA)^2 - \frac{1}{c^2}\left(\frac{\partial S}{\partial t} - e\phi\right)^2 + m_0^2 c^2 = 0, \quad (368)$$

which in Minkowski's notation becomes

$$p_\mu p_\mu = -m_0^2 c^2 = \left(\frac{\partial S}{\partial x_\mu} - eA_\mu\right)\left(\frac{\partial S}{\partial x_\mu} - eA_\mu\right), \quad (369)$$

in a frame of reference in which the initial linear momentum of the electron is zero. The generalized momentum four-vector P_μ is

$$P_\mu := \left(p + eA, \frac{i}{c}(En + e\phi)\right), \quad (370)$$

and is obtained from the sum of four vectors,

174   Chapter 12. $B^{(3)}$ and the Hamilton-Jacobi Equation

$$P_\mu := p_\mu + eA_\mu. \tag{371}$$

As described by Landau and Lifshitz, the fundamental equation of motion (368) is equivalent to the Lagrange equation of motion (where **v** and **r** are non-relativistic)

$$\frac{d}{dt}\left(\frac{\partial L}{\partial \mathbf{v}}\right) = \frac{\partial L}{\partial \mathbf{r}}, \tag{372}$$

where L is the Lagrangian (363). Equation (372) reduces in three dimensional notation to

$$\frac{d\mathbf{p}}{dt} = -e\left(\frac{\partial \mathbf{A}}{\partial t} + \nabla\phi\right) + e\mathbf{v} \times (\nabla \times \mathbf{A}), \tag{373}$$

which is the Lorentz force equation with the well known identities

$$\mathbf{E} := -\frac{\partial \mathbf{A}}{\partial t} - \nabla\phi, \quad \mathbf{B} := \nabla \times \mathbf{A}. \tag{374}$$

In the non-relativistic limit, the Lorentz force equation becomes

$$m_0 \dot{\mathbf{v}} = e(\mathbf{E} + \mathbf{v} \times \mathbf{B}), \tag{375}$$

which, as we shall see, is also the equation of e in A_μ in this limit, provided that **B** is replaced by $\mathbf{B}^{(3)}$, and the time average of **E** is zero.

12.2 SOLUTION OF THE RELATIVISTIC HAMILTON-JACOBI EQUATION (369)

The solution of Eq. (369) is obtained in this section for a circularly polarized plane wave, in which A_μ is a function of the relativistically invariant variable

$$\xi = ct - Z, \tag{376}$$

obtained through the definition

Solution of the Hamilton-Jacobi Equation

$$\xi := -\epsilon_\mu(0, 0, Z, ict) = ct - Z. \tag{377}$$

Thus ξ is identifiable as the electromagnetic phase

$$\xi = -\kappa_\mu x_\mu \tag{378}$$

when the four-vector

$$\kappa_\mu := \left(\boldsymbol{\kappa}, i\frac{\omega}{c}\right) = \kappa\epsilon_\mu \tag{379}$$

is normalized to unit dimensions. The circularly polarized plane wave is therefore characterized, following Landau and Lifshitz [5] by the single variable ξ.

With Eq. (376), the solution proceeds [5] using the Lorentz condition

$$\frac{\partial A_\mu}{\partial x_\mu} = \frac{dA_\mu}{d\xi}\kappa_\mu = 0. \tag{380}$$

The solution of Eq. (369) for a free particle with four-momentum $p_\mu = f_\mu$ takes the form

$$S = f_\mu x_\mu, \tag{381}$$

where

$$f_\mu := (\boldsymbol{f}, if^{(0)}), \qquad f_\mu f_\mu = -m_0^2 c^2. \tag{382}$$

In the presence of electromagnetism, therefore, it is assumed that the general solution takes the form

$$S = f_\mu x_\mu + F(\xi), \tag{383}$$

where $F(\xi)$ is to be determined. Using the assumed solution (383) in Eq. (369) gives

$$2f_\mu \frac{\partial F}{\partial x_\mu}(\xi) - e^2 A_\mu A_\mu - 2eA_\mu f_\mu + \left(\frac{\partial F}{\partial x_\mu}\right)^2 - 2eA_\mu \frac{\partial F}{\partial x_\mu} = 0. \tag{384}$$

Chapter 12. $B^{(3)}$ and the Hamilton-Jacobi Equation

Ignoring the two last quadratic terms, following Landau and Lifshitz [5], leads to the following expression for F,

$$F = \frac{e}{\gamma}\int A_\mu f_\mu d\xi - \frac{e^2}{\gamma}\int A_\mu A_\mu d\xi, \qquad (385)$$

where

$$\gamma := f_\mu \kappa_\mu. \qquad (386)$$

Therefore the action from Eq. (383) becomes the standard solution [5]

$$S = f_\mu x_\mu + \frac{e}{\gamma}\int A_\mu f_\mu d\xi - \frac{e^2}{2\gamma}\int A_\mu A_\mu d\xi \qquad (387)$$

of the equation of motion of e in A_μ.

Conversion to three dimensional notation is achieved [5] with

$$\gamma = f^{(0)} - f_z \qquad (388)$$

and by defining the vector

$$\boldsymbol{\kappa} := f_x \boldsymbol{i} + f_y \boldsymbol{k}, \qquad (389)$$

so that

$$f^{(0)} + f_z = \frac{\left(m_0^2 c^2 + \kappa^2\right)}{\gamma}. \qquad (390)$$

The standard procedure [5] next assumes that $\phi = 0$, and that $\mathbf{A}(\xi)$ is transverse in consequence. If so, the action in three dimensional notation becomes

$$S = \boldsymbol{\kappa}\cdot\boldsymbol{r} - \frac{\gamma}{2}(ct + Z) - \frac{\left(m_0^2 c^2 + \kappa^2\right)\xi}{2\gamma} + \frac{e}{\gamma}\int \mathbf{A}\cdot\boldsymbol{\kappa} d\xi - \frac{e^2}{2\gamma}\int A^2 d\xi. \qquad (391)$$

The momentum vectors of the electron in A_μ are therefore [5]

Solution of the Hamilton-Jacobi Equation

$$P_X = \kappa_X - eA_X, \qquad P_Y = \kappa_Y - eA_Y,$$

$$P_Z = -\frac{\gamma}{2} + \frac{m_0^2 c^2 + \kappa^2}{2\gamma} - \frac{e}{\gamma}\kappa \cdot A + \frac{e^2}{2\gamma}A^2; \tag{392}$$

its energy is

$$En = (\gamma + P_z)c, \tag{393}$$

and its laboratory frame coordinates are

$$X = \gamma^{-1}\left(\kappa_X \xi - e\int A_X d\xi\right), \qquad Y = \gamma^{-1}\left(\kappa_Y \xi - e\int A_Y d\xi\right),$$

$$Z = \frac{1}{2}\left(\frac{m_0^2 c^2 + \kappa^2}{\gamma^2} - 1\right)\xi - \frac{e}{\gamma^2}\int \kappa \cdot A\, d\xi + \frac{e^2}{2\gamma^2}\int A^2 d\xi. \tag{394}$$

Averaging over time [5] in the frame where the electron is initially at rest gives the results

$$P_X = -eA_X, \qquad X = -\frac{e}{\gamma}\int A_X d\xi,$$

$$P_Y = -eA_Y, \qquad Y = -\frac{e}{\gamma}\int A_Y d\xi, \tag{395}$$

$$P_Z = \frac{e^2}{2\gamma}(A^2 - \overline{A}^2), \qquad Z = \frac{e^2}{2\gamma^2}\int (A^2 - \overline{A}^2)\, d\xi.$$

Applying these equations to the real transverse components [5]

$$A_X = -\frac{c}{\omega}B^{(0)}\cos\omega\xi, \qquad A_Y = \frac{c}{\omega}B^{(0)}\sin\omega\xi, \tag{396}$$

gives

$$\overline{A}^2 = A_X^2 + A_Y^2 = \frac{c^2}{\omega^2}B^{(0)2}, \tag{397}$$

and so γ is identifiable as

$$\gamma^2 = \frac{c^2}{\omega^2}\left(m_0^2 \omega^2 + e^2 B^{(0)2}\right). \tag{398}$$

12.3 THE ORBITAL ANGULAR MOMENTUM OF THE ELECTRON IN THE FIELD

The key to the realization of the existence of the field $B^{(3)}$ is that the orbital angular momentum of the electron in the field is a constant, and its Z component is given by

$$J_Z = X p_Y - Y p_X, \qquad (399)$$

where, as in the standard theory of Landau and Lifshitz [5],

$$p_X = \frac{ec}{\omega} B^{(0)} \cos \omega t, \qquad X = -\frac{ec^2 B^{(0)}}{\gamma \omega^2} \sin \omega t,$$

$$p_Y = -\frac{ec}{\omega} B^{(0)} \sin \omega t, \qquad Y = -\frac{ec^2 B^{(0)}}{\gamma \omega^2} \cos \omega t. \qquad (400)$$

Therefore we arrive at the result that the angular momentum component of e in A_μ is

$$J_Z = \frac{e^2 c^2}{\omega^2} \left(\frac{B^{(0)}}{\left(m_0^2 \omega^2 + e^2 B^{(0)2}\right)^{\frac{1}{2}}} \right) B^{(0)}. \qquad (401)$$

The angular momentum vector of e in A_μ is therefore

$$\boldsymbol{J}^{(3)} = \frac{e^2 c^2}{\omega^2} \left(\frac{B^{(0)}}{\left(m_0^2 \omega^2 + e^2 B^{(0)2}\right)^{\frac{1}{2}}} \right) \boldsymbol{B}^{(3)} \qquad (402)$$

and is governed *entirely* by the field $\boldsymbol{B}^{(3)}$.

The electronic angular momentum defines the magnetic dipole moment

$$\boldsymbol{m}^{(3)} = -\frac{e}{2m_0} \boldsymbol{J}^{(3)}, \qquad (403)$$

where $-e/(2m_0)$ is the gyromagnetic ratio and the permanent magnetization is

$$\boldsymbol{M}^{(3)} = N \boldsymbol{m}^{(3)}, \qquad (404)$$

where N is the number of electrons. Therefore the magnetization due to the circularly polarized electromagnetic radia-

Orbital Angular Momentum of the Electron

tion is expressible as

$$\mathbf{M}^{(3)} = \frac{N}{2}(a\chi' + bB^{(0)}\beta'')\mathbf{B}^{(3)}. \qquad (405)$$

Here χ' and β'' are, respectively, the susceptibility and hyperpolarizability

$$\chi' := \frac{-e^2 c^2}{2m_0\omega^2}, \qquad \beta'' := \frac{-e^3 c^2}{2m_0^2\omega^3}, \qquad (406)$$

and the factors a and b are given by

$$a = (1 + x^2)^{-\frac{1}{2}}, \qquad b = \left(1 + \frac{1}{x^2}\right)^{-\frac{1}{2}}, \qquad (407)$$

where

$$x := \frac{m_0\omega}{eB^{(0)}} \qquad (408)$$

is dimensionless in S.I. units.

12.4 LIMITING FORMS OF EQUATION (405)

(a) Under the condition

$$\omega = \frac{e}{m_0} B^{(0)}, \qquad (409)$$

Eq. (405) reduces to

$$\mathbf{M}^{(3)} = \frac{N}{2\sqrt{2}}(\chi' + B^{(0)}\beta'')\mathbf{B}^{(3)}, \qquad (410)$$

which shows conclusively that the magnetization is a sum of terms *linear* in $\mathbf{B}^{(3)}$ and *quadratic* in $\mathbf{B}^{(3)}$. Recall that this is a result of the principle of least action, and is therefore fundamental, not the result of a modelling procedure. In this case the magnetization is a sum of terms to order one half and order one in beam intensity.

(b) Under the condition

$$\omega \leq \frac{e}{m_0} B^{(0)}, \qquad (411)$$

the first term in Eq. (405) predominates and, using a Maclaurin expansion,

$$M^{(3)} \sim \frac{N}{2}\chi'\left(1 + x^2 + \frac{x^4}{2} + \ldots\right)B^{(3)}, \qquad (412)$$

which reduces to the non-relativistic result

$$M^{(3)} \rightarrow \frac{N}{2}\chi' B^{(3)}, \qquad (413)$$

as $\omega \rightarrow 0$, i.e., as the frequency ω becomes very small for a given beam intensity (proportional to $B^{(0)2}$). In this case the magnetization is to order one half in beam intensity.

(c) In the opposite limit,

$$\omega \geq \frac{e}{m_0} B^{(0)}, \qquad (414)$$

the second term in Eq. (405) predominates, and the Maclaurin expansion becomes

$$M^{(3)} \sim \frac{N}{2}\beta''\left(1 + \frac{1}{x^2} + \frac{1}{2x^4} + \ldots\right)B^{(0)}B^{(3)}, \qquad (415)$$

which reduces to

$$M^{(3)} \rightarrow \frac{N}{2}\beta'' B^{(0)} B^{(3)}, \qquad (416)$$

when the frequency ω becomes very large. In this case the magnetization is to order one in beam intensity.

12.5 DISCUSSION

Cases (a), (b) and (c) emerge from the *same* equation of motion of e in A_μ and show conclusively that magnetization by light is governed by the field $\mathbf{B}^{(3)}$. Since ω is the angular frequency of the electron in the beam, $\mathbf{B}^{(3)}$ is the only field property present, and if it were zero, there would

Discussion

be no magnetization at all. As argued in earlier chapters, the field $\boldsymbol{B}^{(3)}$ is a novel fundamental property of electromagnetic radiation. It is noteworthy that Eq. (405) is expressible as

$$\boldsymbol{M}^{(3)} = -i\frac{N}{2}\left(\frac{a\chi'}{B^{(0)}} + b\beta''\right)\boldsymbol{B}^{(1)} \times \boldsymbol{B}^{(2)} \qquad (417)$$

$$= -\frac{iN}{2c^2}\left(\frac{a\chi'}{B^{(0)}} + b\beta''\right)\boldsymbol{E}^{(1)} \times \boldsymbol{E}^{(2)},$$

i.e., in terms of $i\beta''$ and the free space conjugate product

$$\boldsymbol{E}^{(1)} \times \boldsymbol{E}^{(2)} = c^2 \boldsymbol{B}^{(1)} \times \boldsymbol{B}^{(2)}, \qquad (418)$$

proving inter alia that

$$\boldsymbol{B}^{(1)} \times \boldsymbol{B}^{(2)} = iB^{(0)}\boldsymbol{B}^{(3)*} \qquad (419)$$

and that the hyperpolarizability β'' is the imaginary part of a complex property tensor. This is precisely the result obtained in atomic and molecular matter by Woźniak, Evans and Wagnière [18]. If such matter contains net electronic angular momentum, the magnetization term linear in $\boldsymbol{B}^{(3)}$ is also expected, as first discussed by Evans [10]. Magnetization proportional to the light intensity has been observed in the inverse Faraday effect in atoms and molecules [21] as described in Chap. 7. Other magnetic effects of light described there were optical NMR and the optical Faraday effect.

The condition (409) is precisely the non-relativistic limiting condition for the cyclotron frequency, ω, of an electron in a *static magnetic field* [5], and therefore the effect of $\boldsymbol{B}^{(3)}$ on one electron is indistinguishable from that of such a field. This point is discussed further in Appendix D. More generally, the effect of circularly polarized light on a single electron can be explained as in this chapter purely in terms of the magnetic field $\boldsymbol{B}^{(3)}$. Since this effect can be observed routinely in electron plasma [23] it is beyond reasonable doubt that the field $\boldsymbol{B}^{(3)}$ is a physical observable, and therefore a fundamental property of electromagnetic radiation of any frequency and any intensity. Interestingly, the converse also holds true, any static magnetic field ($\boldsymbol{B}_s$) can always be developed mathematically in

182   Chapter 12. $B^{(3)}$ and the Hamilton-Jacobi Equation

terms of a conjugate product, using the reverse of Eq. (419),

$$B_g = -i\frac{B^{(1)} \times B^{(2)}}{B^{(0)}} = B^{(0)}k, \qquad (420)$$

and so on a philosophical level, the concept of conjugate product becomes identified with that of magnetic field, leading to the inference that the antisymmetric part of the light intensity tensor is proportional to a magnetic field. This concept, at first strange and counter intuitive, is nevertheless an inescapable result of the theory of special relativity applied to the equation of motion of e in A_μ at the most fundamental level in the classical theory of fields [5].

Using the condition (411), the experimental conditions can be deduced under which the effect of $B^{(3)}$ can be observed unequivocally. The magnetic flux density amplitude in tesla, $B^{(0)}$, of light (or electromagnetic radiation in general) can be calculated in S.I. units from the standard free space expression

$$B^{(0)} = \left(\frac{W}{\epsilon_0 c^3 Ar}\right)^{\frac{1}{2}} = \left(\frac{I_0}{\epsilon_0 c^3}\right)^{\frac{1}{2}}, \qquad (421)$$

where ϵ_0 is the S.I. free space permittivity ($\epsilon_0 = 8.854 \times 10^{-12} J^{-1} C^2 m^{-1}$), Ar the area of the beam in square meters, and W its power in watts. Therefore I_0 is the power density of the beam in watts per square meter. The ratio e/m_0 is, for the electron,

$$\frac{e}{m_0} = \frac{1.602 \times 10^{-19}}{9.109 \times 10^{-31}} \sim 2 \times 10^{11} \, C\,kgm^{-1}, \qquad (422)$$

and so the angular frequency ω in Eq. (411) must be adjusted to be less than $\sim 2 \times 10^{11} B^{(0)}$ rad s^{-1}. If so, the magnetization of the electron plasma by circularly polarized light is given by Eq. (412), which for very low angular frequency ω reduces to Eq. (413). Under these conditions, the magnetization of the plasma is proportional to the *square root* of the power density of the electromagnetic beam.

The angular frequency ω is the angular velocity of the electron in the beam and from Eq. (42), the effect of $B^{(3)}$ at first order dominates over the effect at second order when an electromagnetic beam is used in the low frequency region (for

Discussion

example, with intense radio frequency beams of high power density directed into an electron plasma). Under the opposite condition (411), achieved with low power densities and high frequencies, the term in $B^{(0)}\boldsymbol{B}^{(3)}$ dominates, and the magnetizing effect of light would be observed to be proportional to light intensity, as discussed in the phenomenological theories of the inverse Faraday effect [10, 18]. Finally, in the condition (409), the observed dependence of magnetization on power density would be a sum of terms in $\boldsymbol{B}^{(3)}$ and in $B^{(0)}\boldsymbol{B}^{(3)}$, as described in Eq. (410).

Experimental observation of these effects is important, because it would demonstrate the existence in free space of the novel field $\boldsymbol{B}^{(3)}$, which as shown in this chapter, is a direct consequence of special relativity. Other important observations include those summarized earlier in this book, namely the inverse and optical Faraday effects due to $\boldsymbol{B}^{(3)}$, optical NMR, (which is in its infancy as a technique), optical ESR, light shifts in atomic spectra due to $\boldsymbol{B}^{(3)}$, the optical Aharonov-Bohm effect, and magnetization in an electron plasma, or electron beam, as described in this chapter. There is little doubt that several other interesting and useful effects await discovery. Nature shows.

It is therefore clear that the theoretical prediction of $\boldsymbol{B}^{(3)}$ opens up a new area of research, centered around the fact that circularly polarized light acts as a magnet upon interaction with material matter.

APPENDIX A. INVARIANCE AND DUALITY IN THE CIRCULAR BASIS

Throughout the text it has been asserted that the novel $\boldsymbol{B}^{(3)}$ is dual to the imaginary $-i\boldsymbol{E}^{(3)}/c$. In this appendix the meaning of this statement is defined using the circular basis (46) for the components of the electromagnetic field tensor $F_{\mu\nu}$ and its dual $G_{\mu\nu}$. Equations (7) of the text are also derived in this basis, showing that the Lorentz invariants L_1 and L_2 of the electromagnetic field in vacuo remain zero in the presence of $\boldsymbol{B}^{(3)}$ and $-i\boldsymbol{E}^{(3)}/c$. If, however, $-i\boldsymbol{E}^{(3)}/c$ is replaced by zero, or is asserted to be real, the Lorentz invariants L_1 and L_2 are no longer zero in vacuo. The assertion that $-i\boldsymbol{E}^{(3)}/c$ is zero is of course contrary to the structure of Eqs. (25) of the text. The assertion that $\boldsymbol{B}^{(3)}$ is zero or imaginary is contrary to the structure of Eqs. (4).

In the basis (46), the complete electric and magnetic field vectors are

$$\boldsymbol{E} = \boldsymbol{E}^{(1)} + \boldsymbol{E}^{(2)} + \boldsymbol{E}^{(3)}, \qquad \boldsymbol{B} = \boldsymbol{B}^{(1)} + \boldsymbol{B}^{(2)} + \boldsymbol{B}^{(3)} \tag{A1}$$

and their complex conjugates are

$$\boldsymbol{E}^* = \boldsymbol{E}^{(1)*} + \boldsymbol{E}^{(2)*} + \boldsymbol{E}^{(3)*}, \qquad \boldsymbol{B}^* = \boldsymbol{B}^{(1)*} + \boldsymbol{B}^{(2)*} + \boldsymbol{B}^{(3)*}. \tag{A2}$$

The individual components are

$$\begin{aligned} \boldsymbol{E}^{(1)} &:= E^{(1)} \boldsymbol{e}^{(1)}, & \boldsymbol{B}^{(1)} &:= B^{(1)} \boldsymbol{e}^{(1)}, \\ \boldsymbol{E}^{(2)} &:= E^{(2)} \boldsymbol{e}^{(2)}, & \boldsymbol{B}^{(2)} &:= B^{(2)} \boldsymbol{e}^{(2)}, \\ \boldsymbol{E}^{(3)} &:= E^{(3)} \boldsymbol{e}^{(3)}, & \boldsymbol{B}^{(3)} &:= B^{(3)} \boldsymbol{e}^{(3)}, \end{aligned} \tag{A3}$$

expressed in terms of the unit vectors $\boldsymbol{e}^{(1)}$, $\boldsymbol{e}^{(2)}$, and $\boldsymbol{e}^{(3)}$ of the basis (46) multiplied by the scalar quantities:

$$\begin{aligned} E^{(1)} &= E^{(0)} e^{i\phi}, & B^{(1)} &= iB^{(0)} e^{i\phi}, \\ E^{(2)} &= E^{(0)} e^{-i\phi}, & B^{(2)} &= -iB^{(0)} e^{-i\phi}, \\ E^{(3)} &= -iE^{(0)}, & B^{(3)} &= B^{(0)}. \end{aligned} \tag{A4}$$

186   Appendix A.

In this basis, the tensors $F_{\mu\nu}$ and $G_{\mu\nu}$ are

$$F_{\mu\nu} = \epsilon_0 \begin{bmatrix} 0 & cB^{(3)} & -cB^{(2)} & -iE^{(1)} \\ -cB^{(3)} & 0 & cB^{(1)} & -iE^{(2)} \\ cB^{(2)} & -cB^{(1)} & 0 & -iE^{(3)} \\ iE^{(1)} & iE^{(2)} & iE^{(3)} & 0 \end{bmatrix} \quad \text{(A5)}$$

and

$$G_{\mu\nu} = \epsilon_0 \begin{bmatrix} 0 & -iE^{(3)} & iE^{(2)} & cB^{(1)} \\ iE^{(3)} & 0 & -iE^{(1)} & cB^{(2)} \\ -iE^{(2)} & iE^{(1)} & 0 & cB^{(3)} \\ -cB^{(1)} & -cB^{(2)} & -cB^{(3)} & 0 \end{bmatrix} \quad \text{(A6)}$$

respectively. The complex conjugate tensors in Minkowski's notation are

$$F^*_{\mu\nu} = \epsilon_0 \begin{bmatrix} 0 & cB^{(3)*} & -cB^{(2)*} & -iE^{(1)*} \\ -cB^{(3)*} & 0 & cB^{(1)*} & -iE^{(2)*} \\ cB^{(2)*} & -cB^{(1)*} & 0 & -iE^{(3)*} \\ iE^{(1)*} & iE^{(2)*} & iE^{(3)*} & 0 \end{bmatrix} \quad \text{(A7)}$$

and

$$G^*_{\mu\nu} = \epsilon_0 \begin{bmatrix} 0 & -iE^{(3)*} & iE^{(2)*} & cB^{(1)*} \\ iE^{(3)*} & 0 & -iE^{(1)*} & cB^{(2)*} \\ -iE^{(2)*} & iE^{(1)*} & 0 & cB^{(3)*} \\ -cB^{(1)*} & -cB^{(2)*} & -cB^{(3)*} & 0 \end{bmatrix}. \quad \text{(A8)}$$

The Lorentz invariants of Eq. (7) of the text are therefore

$$L_1 = F_{\mu\nu} F^*_{\mu\nu} = 0, \qquad L_2 = G_{\mu\nu} G^*_{\mu\nu} = 0. \quad \text{(A9)}$$

These results follow from the fact that the magnitudes of the vectors $\boldsymbol{E}$ and $\boldsymbol{B}$ of Eq. (A1), written in the basis (46) are

$$|\boldsymbol{E}| = (\boldsymbol{E} \cdot \boldsymbol{E}^*)^{\frac{1}{2}}, \qquad |\boldsymbol{B}| = (\boldsymbol{B} \cdot \boldsymbol{B}^*)^{\frac{1}{2}}, \quad \text{(A10)}$$

i.e., are defined in terms of the dot product of the original vector with its complex conjugate, using results such as

$$e^{(1)} \cdot e^{(1)*} = e^{(2)} \cdot e^{(2)*} = e^{(3)} \cdot e^{(3)*} = 1 \tag{A11}$$

for the unit vectors.

It is easily verified that if $B^{(3)}$ is non-zero and real, and if there is no accompanying $E^{(3)}$, the Lorentz invariants L_1 and L_2 are no longer zero.

Using these definitions, the fields $\boldsymbol{E}^{(i)}$ are dual to $ic\boldsymbol{B}^{(i)}$, where i = 1, 2, and 3. Similarly, the conjugate fields $\boldsymbol{E}^{(i)*}$ are dual to $ic\boldsymbol{B}^{(i)*}$. We see that the component $iE^{(3)}/c$ is dual to $B^{(3)}$, and if we define a vector $\boldsymbol{E}^{(3)}$ whose magnitude is $E^{(3)}$, then the vector $-i\boldsymbol{E}^{(3)}/c$ is dual to $\boldsymbol{B}^{(3)}$. This terminology is arrived at from the mathematical theorem that $F_{\mu\nu}$ is dual to $G_{\mu\nu}$. Thus, if $\boldsymbol{B}^{(3)}$ is real, then the longitudinal electric field component is imaginary, an inference which is also supported by the fact that there is experimental evidence (Chap. 7) for $\boldsymbol{B}^{(3)}$, and none for a putative real $\boldsymbol{E}^{(3)}$. We have seen in Chap. 2, finally, that discrete symmetry is violated if a cyclical relation is constructed using a real $\boldsymbol{E}^{(3)}$.

APPENDIX B. ANGULAR MOMENTUM IN SPECIAL RELATIVITY

In special relativity, angular momentum is represented by an antisymmetric four-tensor of the form

$$J_{\mu\nu} = x_\mu p_\nu - x_\nu p_\mu, \qquad (B1)$$

where x_μ and p_μ are both four-vectors. The symmetry of the angular momentum tensor is the same as that of the electromagnetic four-tensor $F_{\mu\nu}$. Following Barut [46] the angular momentum four-tensor can be written as

$$J_{\rho\sigma} = \begin{bmatrix} 0 & J_Z & -J_Y & -iJ_{10} \\ -J_Z & 0 & J_X & -iJ_{20} \\ J_Y & -J_X & 0 & -iJ_{30} \\ iJ_{10} & iJ_{20} & iJ_{30} & 0 \end{bmatrix}, \qquad (B2)$$

where the space-like elements are interpreted as in Newtonian theory. The other elements are defined as

$$J_{k0} = x_k p_0 - p_k x_0. \qquad (B3)$$

The dual of the angular momentum tensor is therefore

$$J_{\mu\nu}^{(D)} = \frac{1}{2}\epsilon_{\mu\nu\rho\sigma} J_{\rho\sigma} = \begin{bmatrix} 0 & -iJ_{30} & iJ_{20} & J_X \\ iJ_{30} & 0 & -iJ_{10} & J_Y \\ -iJ_{20} & iJ_{10} & 0 & J_Z \\ -J_X & -J_Y & -J_Z & 0 \end{bmatrix}. \qquad (B4)$$

The Pauli-Lubanski pseudo-vector corresponding to the tensor (B2) is defined, following Ryder [54] as

$$J_\mu := \frac{-i}{2}\epsilon_{\mu\nu\rho\sigma} J_{\rho\sigma} p_\nu = -i J_{\mu\nu}^{(D)}$$

$$= (J_{20} + J_X, -J_{10} + J_Y, J_Z, iJ_Z). \qquad (B5)$$

In general, this pseudo four-vector has four components, and is formed from the dual tensor (B4) by multiplication with the linear momentum four-vector p_ν [54].

It is well known that the helicity of the free photon is defined through the fact [54] that its angular momentum four-vector is directly proportional to its linear momentum four-vector. This means that the J_μ vector of the photon must take the form

$$J_\mu = (0, 0, J_Z, iJ_Z) = \hbar(0, 0, 1, i), \qquad (B6)$$

because its linear momentum (Chap. 11) is

$$p_\mu = \hbar\kappa(0, 0, 1, i). \qquad (B7)$$

It follows that for the free photon, two of the space-like components of J_μ vanish. If the photon propagates in Z, the X and Y components vanish, meaning that

$$J_{20} + J_X = 0, \qquad -J_{10} + J_Y = 0. \qquad (B8)$$

We have seen in the text that rotation generators become angular momentum operators in quantum mechanics, and are proportional to magnetic field operators. Similarly, electric field operators are proportional to boost generators. It therefore becomes clear that Eqs. (B8) of this appendix are paralleled by the equations of free space electrodynamics

$$E_Y + cB_X = 0, \qquad -E_X + cB_Y = 0. \qquad (B9)$$

The angular momentum component J_X is proportional to the free space magnetic field component B_X, and the component J_{20} to the free space electric field component E_Y. Therefore J_X is proportional to a rotation generator, and J_{20} to a boost generator.

Finally, we note that in Minkowski's notation, it is necessary to define

$$J_\mu = \frac{-i}{2} \epsilon_{\mu\nu\rho\sigma} J_{\rho\sigma} p_\nu, \qquad (B10)$$

in order to retrieve a real Pauli-Lubansky four-vector J_μ. The factor $-i$ premultiplying the left hand side is a consequence of Minkowski's representation of space-time. Similarly, the factor $-i$ is needed in Eq. (342) of Chap. 11 to ensure that B_z, which is directly proportional to the photon angular momentum, is real, as indicated by Eqs. (4). The use of Minkowski's notation, rather than covariant-contravariant notation, makes it clear that if B_z is real, its dual $-iE_z/c$ is imaginary, and unphysical, although non-zero.

In conclusion therefore, the retrieval of a real photon angular momentum from $J_{\mu\nu}$ exactly parallels the retrieval of a real B_z from the four-tensor $F_{\mu\nu}$. Thus if J_z is physical, so is B_z, and this is precisely the result embodied in Eq. (5) of the text.

APPENDIX C. STANDARD EXPRESSIONS FOR THE ELECTROMAGNETIC FIELD IN FREE SPACE, WITH LONGITUDINAL COMPONENTS

Maxwell's equations remain invariant under the duality transformation,

$$E^{(0)} \rightarrow icB^{(0)}, \quad (C1)$$

which means that the Cartesian components of the tensor $F_{\mu\nu}$ transform as

$$E_X \rightarrow icB_X, \quad B_X \rightarrow \frac{-iE_X}{c},$$

$$E_Y \rightarrow icB_Y, \quad B_Y \rightarrow \frac{-iE_Y}{c}, \quad (C2)$$

$$E_Z \rightarrow icB_Z, \quad B_Z \rightarrow \frac{-iE_Z}{c}.$$

The complete representation of the free space electromagnetic field in one sense of circular polarization is as follows,

$$\boldsymbol{E} = \boldsymbol{E}^{(1)} + \boldsymbol{E}^{(2)} + \boldsymbol{E}^{(3)}, \quad \boldsymbol{B} = \boldsymbol{B}^{(1)} + \boldsymbol{B}^{(2)} + \boldsymbol{B}^{(3)}, \quad (C3)$$

and each circular component can be represented in terms of the Cartesian components of $F_{\mu\nu}$.

C.1 COMPONENT (1)

$$\boldsymbol{E}^{(1)} = E_X^{(1)} \boldsymbol{i} + iE_Y^{(1)} \boldsymbol{j}, \quad \boldsymbol{B}^{(1)} = iB_X^{(1)} \boldsymbol{i} + B_Y^{(1)} \boldsymbol{j}, \quad (C4)$$

where $E_X^{(1)} = E^{(0)}/\sqrt{2}\, e^{i\phi}$, $E_Y^{(1)} = -E^{(0)}/\sqrt{2}\, e^{i\phi}$, $B_X^{(1)} = B^{(0)}/\sqrt{2}\, e^{i\phi}$, $B_Y^{(1)} = B^{(0)}/\sqrt{2}\, e^{i\phi}$. Switching from one sense of circular polarization to the other results in

$$E_Y^{(1)} \rightarrow -E_Y^{(1)}, \quad B_X^{(1)} \rightarrow -B_X^{(1)}. \quad (C5)$$

The $F_{\mu\nu}$ matrix for component (1) is therefore

$$F_{\mu\nu}^{(1)} = \epsilon_0 \begin{bmatrix} 0 & 0 & -cB_Y^{(1)} & -iE_X^{(1)} \\ 0 & 0 & cB_X^{(1)} & -iE_Y^{(1)} \\ cB_Y^{(1)} & -cB_X^{(1)} & 0 & 0 \\ iE_X^{(1)} & iE_Y^{(1)} & 0 & 0 \end{bmatrix}, \qquad (C6)$$

and the duality transformation (C1) takes this to the $G_{\mu\nu}^{(1)}$ matrix, an operation corresponding to

$$G_{\mu\nu}^{(1)} = \frac{1}{2} \epsilon_{\mu\nu\rho\sigma} F_{\rho\sigma}^{(1)}. \qquad (C7)$$

C.2 COMPONENT (2)

Similarly, for component (2), the $\boldsymbol{E}^{(2)}$ and $\boldsymbol{B}^{(2)}$ fields are

$$\boldsymbol{E}^{(2)} = E_X^{(2)} \boldsymbol{1} - iE_Y^{(2)} \boldsymbol{j}, \qquad \boldsymbol{B}^{(2)} = -iB_X^{(2)} \boldsymbol{1} + B_Y^{(2)} \boldsymbol{j}, \qquad (C8)$$

where $E_X^{(2)} = E^{(0)}/\sqrt{2}\, e^{-i\phi}$, $E_Y^{(2)} = -E^{(0)}/\sqrt{2}\, e^{-i\phi}$, $B_X^{(2)} = B^{(0)}/\sqrt{2}\, e^{-i\phi}$, $B_Y^{(2)} = B^{(0)}/\sqrt{2}\, e^{-i\phi}$ and the matrix $F_{\mu\nu}^{(2)}$ corresponding to component (2) is, in free space

$$F_{\mu\nu}^{(2)} = F_{\mu\nu}^{(1)*} = \epsilon_0 \begin{bmatrix} 0 & 0 & -cB_Y^{(2)} & -iE_X^{(2)} \\ 0 & 0 & cB_X^{(2)} & -iE_Y^{(2)} \\ cB_Y^{(2)} & -cB_X^{(2)} & 0 & 0 \\ iE_X^{(2)} & iE_Y^{(2)} & 0 & 0 \end{bmatrix}. \qquad (C9)$$

Switching from one sense of circular polarization to the other for component (2) results in,

$$E_Y^{(2)} \to -E_Y^{(2)}, \qquad B_X^{(2)} \to -B_X^{(2)}. \qquad (C10)$$

C.3 COMPONENT (3)

The longitudinal magnetic component (3) is pure real and is given by

The Electromagnetic Field in Free Space

$$\mathbf{B}^{(3)} = B^{(0)}\mathbf{k} = B_Z^{(3)}\mathbf{k}, \tag{C11}$$

so that the duality transformation (C1) gives the pure imaginary longitudinal electric component (3),

$$\mathbf{E}^{(3)} = -iE^{(0)}\mathbf{k} = -iE_Z^{(3)}\mathbf{k}. \tag{C12}$$

Switching the sense of circular polarization in this case results in

$$B_Z^{(3)} \to -B_Z^{(3)}, \qquad E_Z^{(3)} \to -E_Z^{(3)}, \tag{C13}$$

and the $F_{\mu\nu}^{(3)}$ matrix for component (3) is

$$F_{\mu\nu}^{(3)} = \epsilon_0 \begin{bmatrix} 0 & cB_Z^{(3)} & 0 & 0 \\ -cB_Z^{(3)} & 0 & 0 & 0 \\ 0 & 0 & 0 & -iE_Z^{(3)} \\ 0 & 0 & iE_Z^{(3)} & 0 \end{bmatrix}, \tag{C14}$$

the dual of which in free space is $G_{\mu\nu}^{(3)}$,

$$G_{\mu\nu}^{(3)} = \epsilon_0 \begin{bmatrix} 0 & -iE_Z^{(3)} & 0 & 0 \\ iE_Z^{(3)} & 0 & 0 & 0 \\ 0 & 0 & 0 & cB_Z^{(3)} \\ 0 & 0 & -cB_Z^{(3)} & 0 \end{bmatrix}. \tag{C15}$$

Note that both in the $F_{\mu\nu}^{(3)}$ matrix and in the $G_{\mu\nu}^{(3)}$ matrix there appear a real $B_Z^{(3)}$ and an imaginary $-iE_Z^{(3)}$.

The pseudo four-vector dual to the $F_{\rho\sigma}^{(3)}$ matrix is (Chap. 11),

$$V_\mu^{(3)} = \frac{-i}{2} \epsilon_{\mu\nu\rho\sigma} F_{\rho\sigma}^{(3)} \epsilon_\nu = -iG_{\mu\nu}^{(3)} \epsilon_\nu, \tag{C16}$$

which is $V_\mu^{(3)} = \epsilon_0 c(0, 0, B_Z, iB_Z)$. Similarly $W_\mu^{(3)}$ is dual to $G_{\rho\sigma}^{(3)}$ as in Chap. 11. It is important to note that *in free space* the pseudo four-vectors, $V_\mu^{(1)}$ and $V_\mu^{(2)}$, dual to the four-tensors $F_{\rho\sigma}^{(1)}$ and $F_{\rho\sigma}^{(2)}$ respectively, are both null,

$$V_\mu^{(1)} = -iG_{\mu\nu}^{(1)}\epsilon_\nu = \epsilon_0\left(E_Y^{(1)} + cB_X^{(1)}, -E_X^{(1)} + cB_Y^{(1)}, 0, 0\right) = 0,$$
$$V_\mu^{(2)} = -iG_{\mu\nu}^{(2)}\epsilon_\nu = \epsilon_0\left(E_Y^{(2)} + cB_X^{(2)}, -E_X^{(2)} + cB_Y^{(2)}, 0, 0\right) = 0,$$
(C17)

because

$$E_Y^{(1)} = -cB_X^{(1)}, \qquad E_X^{(1)} = cB_Y^{(1)}, \qquad (C18)$$

and similarly for component (2). Therefore the only non-zero pseudo four-vector is $V_\mu^{(3)}$, which means that the angular momentum of the photon in free space must be directly proportional to $\boldsymbol{B}^{(3)}$, as in the text, and that the helicity of the photon is directly proportional to $\boldsymbol{B}^{(3)}$, as discussed in Chap. 11.

Therefore if $\boldsymbol{B}^{(3)}$ is asserted to be zero, the angular momentum and helicity of the photon both vanish, an incorrect result. Using this reduction to absurdity it becomes clear that for self consistency, the theory of electromagnetism in vacuo must produce a non-zero $\boldsymbol{B}^{(3)}$, as in Eq. (4) of the text.

The Maxwell equations produce the result (C19) only in free space and therefore $V_\mu^{(1)}$ and $V_\mu^{(2)}$ are null only in free space. In general there exist non-zero pseudo four-vectors, whose components are electric and magnetic field components. These pseudo four-vectors do not transform in the same way as ordinary four-vectors under the Lorentz transformation, their individual components transform as components of the anti-symmetric four-tensor dual to the pseudo four-vector. Therefore pseudo four-vectors are four dimensional equivalents of pseudo-vectors in three dimensional space.

C.4 RELATION TO BOOST AND ROTATION GENERATORS

The electromagnetic four-tensors $F_{\mu\nu}^{(1)}$, $F_{\mu\nu}^{(2)}$ and $F_{\mu\nu}^{(3)}$ can be expressed in terms of the boost and rotation generators of the Lorentz group [54]. For example, $F_{\mu\nu}^{(3)}$ is a combination of the longitudinal rotation generator $J_{\mu\nu}^{(3)}$ and longitudinal boost generator $K_{\mu\nu}^{(3)}$, where [54]

$$J_{\mu\nu}^{(3)} = \begin{bmatrix} 0 & -i & 0 & 0 \\ i & 0 & 0 & 0 \\ 0 & 0 & 0 & 0 \\ 0 & 0 & 0 & 0 \end{bmatrix}, \quad K_{\mu\nu}^{(3)} = \begin{bmatrix} 0 & 0 & 0 & 0 \\ 0 & 0 & 0 & 0 \\ 0 & 0 & 0 & 1 \\ 0 & 0 & -1 & 0 \end{bmatrix}. \quad (C19)$$

The Electromagnetic Field in Free Space

Therefore

$$F_{\mu\nu}^{(3)} = \epsilon_0 c B^{(0)} \left(J_{\mu\nu}^{(3)} + K_{\mu\nu}^{(3)} \right), \tag{C20}$$

and the dual pseudo four-vector can be expressed as

$$V_{\mu}^{(3)} = \frac{1}{2} \epsilon_0 c B^{(0)} \epsilon_{\mu\nu\rho\sigma} \left(J_{\rho\sigma}^{(3)} + K_{\rho\sigma}^{(3)} \right) \epsilon_{\nu}, \tag{C21}$$

with similar expressions for $V_{\mu}^{(1)}$ and $V_{\mu}^{(2)}$ in terms of the other boost and rotation generators. Equation (C21) shows that the pseudo four-vector $V_{\mu}^{(3)}$ is defined within the structure of the inhomogeneous Lorentz group (or Poincaré group) with the use of the unit photon linear momentum four-vector. The latter is proportional to the generator of translations in space-time, and as first shown by Wigner, the photon helicity is the ratio of the Pauli-Lubansky pseudo four-vector of photon angular momentum to the generator of space-time translations [51]. Photon helicity can be defined, therefore, only within the Poincaré group, and cannot be defined within the Lorentz group. A similar deduction accrues for $B^{(3)}$ which must be defined in terms of space-time translation.

APPENDIX D. THE LORENTZ FORCE DUE TO $F_{\mu\nu}^{(3)}$; AND $T_{\mu\nu}^{(3)}$

In S.I. units, the Lorentz force due to $F_{\mu\nu}^{(3)}$ is given by

$$f_\mu^{(3)} = F_{\mu\nu}^{(3)} j_\nu, \tag{D1}$$

where the current four-vector is defined as

$$j_\nu = e\left(\frac{v_X}{c}, \frac{v_Y}{c}, \frac{v_Z}{c}, i\right). \tag{D2}$$

Equation (D1) gives the result

$$f_\mu^{(3)} = e\left(B_Z v_Y, -B_Z v_X, E_Z, iE_Z \frac{v_Z}{c}\right). \tag{D3}$$

However, the longitudinal electric field is *defined* as

$$\mathbf{E}^{(3)} := -iE_Z \mathbf{k}, \tag{D4}$$

so that the physical part of the force four-vector is

$$f_\mu^{(3)} = e(B_Z v_Y, -B_Z v_X, 0, 0). \tag{D5}$$

which in three dimensions becomes

$$\mathbf{f}^{(3)} = e\mathbf{v} \times \mathbf{B}^{(3)}. \tag{D6}$$

This means that the magnetic field $\mathbf{B}^{(3)}$ drives an electron in a circle, so that the effect of $\mathbf{B}^{(3)}$ on one electron is qualitatively the same as that of the transverse components. In other words the transverse and longitudinal components of a circularly polarized electromagnetic field both drive an electron in a circular orbit. In the first part of this appendix, we derive the condition under which this orbit is quantitatively the same, i.e., under which the transverse momenta imparted to the electron by $B^{(3)}$ and by the transverse electromagnetic fields are identical.

The transverse electron momentum from the ordinary transverse, oscillating components of the electromagnetic field is given by

$$P_\perp = \left(P_X^2 + P_Y^2\right)^{\frac{1}{2}}, \tag{D7}$$

where the X and Y momenta are defined from

$$P_X = -eA_X^{(1)}, \quad P_Y = -eA_Y^{(1)}, \tag{D8}$$

where

$$\mathbf{A}^{(1)} = \frac{B^{(0)}c}{\sqrt{2}\omega}(i\mathbf{1} + \mathbf{j})e^{i\phi} \tag{D9}$$

is the transverse vector potential of the field and e the electronic charge. We obtain

$$P_\perp = \frac{ecB^{(0)}}{\omega}, \tag{D10}$$

a well known result in elementary field theory.

If the vector potential of $B^{(3)}$ is given by (Chap. 10)

$$\mathbf{A}_3 = \frac{B^{(0)}c}{2}(-Y\mathbf{1} + X\mathbf{j}), \tag{D11}$$

a similar calculation shows that the electronic transverse momentum due to $B^{(3)}$ is

$$P_\perp = eB^{(0)}\frac{R}{2}, \quad R = (X^2 + Y^2)^{\frac{1}{2}}. \tag{D12}$$

Therefore the momenta (D10) and (D12) are the same under the condition

$$a := \frac{R}{2} = \frac{c}{\omega} = \frac{\lambda}{2\pi}, \tag{D13}$$

where a is the gyration radius of the electron and λ the wavelength of the transverse part of the electromagnetic field. The condition (D13) is a constraint on the variables

Lorentz Force due to $F_{\mu\nu}^{(3)}$; and $T_{\mu\nu}^{(3)}$

X and Y of the vector potential of $B^{(3)}$ showing that when the gyration radius of the electron under the influence of $B^{(3)}$ is c/ω, its orbit is *precisely the same* under the influence of $B^{(3)}$ and under the influence of the transverse (oscillating) field components.

Furthermore, the constraint (D13) can be used to define the vector potential of $B^{(3)}$ under condition (D12),

$$A_3 = \frac{B^{(0)}c}{2}(-Y\mathbf{i} + X\mathbf{j}), \quad (X^2 + Y^2)^{\frac{1}{2}} = \frac{2c}{\omega} = \frac{\lambda}{\pi}, \tag{D14}$$

and therefore to define $B^{(3)}$ itself. We again conclude therefore, that the presence of oscillating, transverse, electromagnetic field components in free space implies the presence of $B^{(3)}$ in free space, as in Eqs. (4) of Chap. 1. Recall that $B^{(3)}$ is defined by the experimentally measurable conjugate product,

$$B^{(1)} \times B^{(2)} = iB^{(0)}B^{(3)*}, \tag{D15}$$

and although the angular frequency (ω) does not appear *explicitly* in $B^{(3)}$, it is implicit in its definition, and appears, therefore, in the definition of the vector potential due to $B^{(3)}$, i.e., it appears in Eq.(D14) as demonstrated. The conjugate product is experimentally measurable, so $B^{(3)}$ is physical, and therefore so is A_3. The latter should therefore cause an optical Aharonov-Bohm effect, as in Chap. 9.

D.1 THE STRESS-ENERGY-LINEAR MOMENTUM TENSOR $T_{\mu\nu}^{(3)}$

Using a combination of the Lorentz force equation and the inhomogenous Maxwell equations it is well known [4, 45] that the Lorentz force can be expressed as

$$f_\mu = \frac{1}{\epsilon_0^2} F_{\mu\nu} \frac{\partial F_{\nu\lambda}}{\partial x_\lambda} = \frac{1}{\epsilon_0} \frac{\partial T_{\mu\nu}}{\partial x_\nu}, \tag{D16}$$

where (in S.I. units) the stress-energy-momentum tensor is

$$T_{\mu\nu} = \frac{1}{\epsilon_0}\left(F_{\mu\lambda}F_{\lambda\nu} + \frac{1}{4}\delta_{\mu\nu}F_{\lambda\sigma}F_{\lambda\sigma}\right), \tag{D17}$$

an expression first derived by Minkowski. In free space, the Lorentz force is taken to be zero, so the four-derivative of $T_{\mu\nu}$ vanishes,

$$\frac{\partial T_{\mu\nu}}{\partial x_\nu} = 0. \tag{D18}$$

In free space, the tensor $T_{\mu\nu}^{(1)}$ defined from $F_{\mu\nu}^{(1)}$ by Eq.(D17), is

$$T_{\mu\nu}^{(1)} = \epsilon_0 \begin{bmatrix} 0 & 0 & 0 & 0 \\ 0 & 0 & 0 & 0 \\ 0 & 0 & -c^2 B^{(0)2} & -iE^{(0)}B^{(0)}c \\ 0 & 0 & -iE^{(0)}B^{(0)}c & E^{(0)2} \end{bmatrix} e^{2i\phi} \tag{D19}$$

and the tensor $T_{\mu\nu}^{(3)}$ defined from $F_{\mu\nu}^{(3)}$ is

$$T_{\mu\nu}^{(3)} = \epsilon_0 \begin{bmatrix} -c^2 B^{(0)2} & 0 & 0 & 0 \\ 0 & -c^2 B^{(0)2} & 0 & 0 \\ 0 & 0 & E^{(0)2} & 0 \\ 0 & 0 & 0 & E^{(0)2} \end{bmatrix}. \tag{D20}$$

In each case, the (4,4) element is the electromagnetic energy [4, 45], and the (i,4) and (4,i) elements, i = 1, 2, 3, are components of the electromagnetic linear momentum, proportional to the Poynting vector. Therefore $T_{\mu\nu}^{(3)}$ has no linear electromagnetic momentum, because it is diagonal, whereas $T_{\mu\nu}^{(1)}$ (and $T_{\mu\nu}^{(2)}$) both generate linear electromagnetic momentum. This is consistent with the fact that the Poynting vector $-\boldsymbol{B}^{(3)} \times \boldsymbol{E}^{(3)}/c$ is always zero, $\boldsymbol{B}^{(3)}$ being parallel to $-i\boldsymbol{E}^{(3)}/c$. Note also that $T_{\mu\nu}^{(3)}$, being diagonal, generates no off-diagonal Maxwell stresses in free space, but retains the property [4, 45]

$$T_{\mu\mu}^{(3)} = 0, \tag{D21}$$

i.e., the sum of diagonals vanishes in free space.

An insightful result is obtained by integrating Eq.(D18) in free space, described as follows. The relativistically correct integral is [45]

Lorentz Force due to $F^{(3)}_{\mu\nu}$; and $T^{(3)}_{\mu\nu}$

$$G_\mu = \int p_\nu T_{\mu\nu} dV_0, \qquad (D22)$$

where

$$J_\mu = p_\mu V_0, \qquad (D23)$$

dV_0 being the infinitesimal of proper volume, where V_0 is a Lorentz invariant. Here p_μ is the four-momentum. The latter is v_μ/m_0 where v_μ is the four-velocity and m_0 the relativistically invariant rest mass. Therefore a relativistically correct (i.e., correctly covariant) description of the four-momentum generated by $T_{\mu\nu}$ is given by the integral

$$p_\nu = m_0 v_\nu = \frac{1}{c} \int T_{\mu\nu} \epsilon_\nu dV_0, \qquad (D24)$$

where ϵ_ν is the unit four-momentum defined in Chap. 11. For a particle travelling at the speed of light

$$\epsilon_\nu = (0, 0, 1, i). \qquad (D25)$$

That Eq.(D24) has the correct units can be seen from the fact that $T_{\mu\nu}/c$ has the units of momentum density (momentum per unit volume), and p_ν has the units of momentum. Here ϵ_ν is unitless and V_0 is of course a relativistically invariant rest-volume.

Using ϵ_ν leads to the covariant definition of volume,

$$V_\mu = \epsilon_\mu V_0, \qquad (D26)$$

for a particle travelling at the speed of light. Working out $T^{(3)}_{\mu\nu} \epsilon_\nu$ in Eq.(D24) leads to the familiar result

$$\hbar\omega = \hbar\kappa c = \epsilon_0 \int E^{(0)2} dV_0 \qquad (D27)$$

for the energy-momentum four-vector of a photon. The remarkable fact about this simple calculation is that

$$T^{(1)}_{\mu\nu} \epsilon_\nu = T^{(2)}_{\mu\nu} \epsilon_\nu = 0, \qquad (D28)$$

so that the energy-momentum from the transverse $T^{(1)}_{\mu\nu}$ (or $T^{(2)}_{\mu\nu}$) cannot be defined. The only way of defining it in a relativistically correct way is to use the longitudinal tensor $T^{(3)}_{\mu\nu}$, which as we have seen, is diagonal. Equation (D24) also allows for the fact that there may be non-zero mass present in radiation, because, as in standard special relativity [46] the four-momentum is the product of rest mass and four-velocity,

$$p_\mu = m_0 v_\mu. \tag{D29}$$

As ably described by Barut [46] the concept of a massless particle does not exist in Newtonian dynamics, so a concept such as p_μ has no non-relativistic counterpart. The relation (D29) has no meaning if the mass is zero, and as described in the text, energy and momentum become primary concepts when dealing with the hypothetical massless photon, but in this case we lose the causal connection between momentum and velocity, and localization of the massless photon becomes meaningless. Therefore the unit four-momentum p_μ can be defined only in special relativity, using the quantum hypothesis [46]

$$En = \hbar\omega, \quad \mathbf{p} = \hbar\mathbf{\kappa}, \tag{D30}$$

a hypothesis which is still valid when there is finite mass. However, for finite mass, the wave four-vector

$$\kappa_\mu = \left(\mathbf{\kappa}, i\frac{\omega}{c}\right), \tag{D31}$$

is no longer light-like, and we obtain de Broglie's matter waves, and de Broglie's Guiding theorem, Eq.(1) of the text. For the massive photon, the unit four-vector ϵ_μ is no longer light-like, and p_μ can be related to v_μ in Eq.(D29) with a finite m_0. For finite m_0, there is a photon rest frame, but for zero m_0, there is no photon rest frame, resulting in the familiar idea that the photon is massless because it travels at the speed of light.

REFERENCES

(1) for example, *American Heritage Dictionary*.
(2) J.-P. Vigier, "Present Experimental Status of the Einstein-de Broglie Theory of Light.", *Proceedings of the I.S.Q.M. 1992 Workshop on Quantum Mechanics* (Tokyo, 1992).
(3) L. de Broglie, *Méchanique Ondulatoire du Photon, et Theorie Quantique des Champs* (Gauthier-Villars, Paris, 1957).
(4) J. D. Jackson, *Classical Electrodynamics* (Wiley, New York, 1962).
(5) L. D. Landau and E. M. Lifshitz, *The Classical Theory of Fields*, 4th edn. (Pergamon, Oxford, 1975).
(6) M. Born and E. Wolf, *Principles of Optics*, 6th edn. (Pergamon, Oxford, 1975).
(7) S. Kielich in *Dielectric and Related Molecular Processes*, Vol. 1 (Chemical Society, London, 1972).
(8) M. Whitner, *Electromagnetics* (Prentice Hall, Englewood Cliffs, 1962).
(9) M. W. Evans, *Physica B* **182**, 227 (1992).
(10) M. W. Evans, ibid., **182**, 237 (1992); **183**, 103 (1993).
(11) M. W. Evans, in *Waves and Particles in Light and Matter*, A Garuccio and A. van der Merwe, eds. (Plenum, New York, 1994).
(12) M. W. Evans, *The Photon's Magnetic Field* (World Scientific, Singapore, 1992).
(13) A. A. Hasanein and M. W. Evans, *Quantum Chemistry and the Photomagneton* (World Scientific, Singapore, 1994).
(14) M. W. Evans, *Mod. Phys. Lett.* **7**, 1247 (1993); *Found. Phys. Lett.* **7**, 67 (1994); *Found. Phys.*, in press.
(15) M. W. Evans and S. Kielich eds., *Modern Nonlinear Optics*, Vols. **85**(1), **85**(2), **85**(3) of *Advances in Chemical Physics*, I. Prigogine and S. A. Rice, eds. (Wiley Interscience, New York, 1993/1994). Volume **85**(2) contains a discussion of the cyclic algebra.
(16) W. S. Warren, S. Mayr, D. Goswami, and A. P. West Jr., *Science* **255**, 1683 (1992); **259**, 836 (1993).
(17) R. A. Harris and I. Tinoco, *Science* **259**, 835 (1993).
(18) S. Woźniak, M. W. Evans, and G. Wagnière, *Mol. Phys.* **75**, 81, 99 (1992).

(19) A. Piekara and S. Kielich, *Arch. Sci.* **11**, 304 (1958); *Acta Phys. Pol.* **32**, 405 (1967); theoretical predictions verified experimentally by van der Ziel et al. (Ref. 21), also predicted theoretically by Pershan, Ref. 20.
(20) P. S. Pershan, *Phys. Rev.* **130**, 919 (1963).
(21) J. P. van der Ziel, P. S. Pershan, and L. D. Malmstrom, *Phys. Rev. Lett.* **15**, 190 (1965); also *Phys. Rev.* **143**, 574 (1966).
(22) P. W. Atkins and M. H. Miller, *Mol. Phys.* **75**, 491, 503 (1968); these papers predict theoretically both an inverse and optical Faraday effect, the latter was observed experimentally by Sanford et al. (Ref. 25). The history of the IFE and OFE is reviewed by Zawodny in Vol. **85**(1) of Ref. 15.
(23) J. Deschamps, M. Fitaire, and M. Lagoutte, *Phys. Rev. Lett.* **25**, 1330 (1970); also *Rev. Appl. Phys.* **7**, 155 (1972).
(24) T. W. Barrett, H. Wohltjen, and A. Snow, *Nature* **301**, 694 (1983).
(25) N. Sanford, R. W. Davies, A. Lempicki, W. J. Miniscalco, and S. J. Nettel, *Phys. Rev. Lett.* **50**, 1803 (1983).
(26) S. Woźniak, G. Wagnière and R. Zawodny, *Phys. Lett. A* **154**, 259 (1991); see also Ref. 18 for a detailed development.
(27) A. Pais, *The Science and The Life of Albert Einstein* (Oxford University Press, New York, 1982).
(28) A. Einstein, *Ann. Phys.* **17**, 132 (1905).
(29) reviewed recently by J.-P. Vigier, Ref. 2.
(30) A. S. Goldhaber and M. M. Nieto, *Rev. Mod. Phys.* **43**(3), 277 (1971).
(31) D. Bohm and J.-P. Vigier, *Phys. Rev.* **109**, 1882 (1958); L. de Broglie and J.-P. Vigier, *Phys. Rev. Lett.* **28**, 1001 (1972).
(32) J.-P. Vigier, *I.E.E.E. Trans. Plasma Sci.* **18**, 64 (1990).
(33) M. W. Evans and J.-P. Vigier, *Found. Phys.* in press.
(34) J.-P. Vigier, "Real Physical Paths in the Quantum Mechanical Equivalence of the Einstein-de Broglie and Feynman Points of View in Quantum Mechanics.", *3rd. Int. Symp. Quant. Mech.* Tokyo, 140 (1989).
(35) J.-P. Vigier, *Found. Phys.* **21**, 125 (1991).
(36) for example, D. E. Soper, *Classical Field Theory* (Wiley, New York, 1976).

References

(37) E. Noether, *Nachr. Ges. Wiss. Göttingen* **171** (1918); reviewed by J. G. Fletcher, *Rev. Mod. Phys.* **32**, 65 (1960).
(38) for example, P. W. Atkins, *Molecular Quantum Mechanics* (Oxford University Press, Oxford, 1983).
(39) R. M. Whitner, *Electromagnetics* (Prentice Hall, Englewood Cliffs, 1962).
(40) S. Kielich, *Molecular Nonlinear Optics* (Nauka, Moscow, 1981).
(41) J. L. Synge, *Relativity: The Special Theory* (North Holland, Amsterdam, 1956).
(42) J. Aharoni, *The Special Theory of Relativity* (Clarendon, Oxford, 1959).
(43) given in numerous textbooks in electrodynamics and field theory. Reference 4 is considered a classic text.
(44) D. Corson and P. Lorrain, *Introduction to Electromagnetic Waves and Fields* (Freeman, San Francisco, 1962).
(45) W. K. H. Panofsky and M. Phillips, *Classical Electricity and Magnetism*, 2nd edn. (Addison Wesley, Reading, 1962).
(46) A. O. Barut, *Electrodynamics and Classical Theory of Fields and Particles* (Macmillan, New York, 1964).
(47) L. de Broglie, *Annales de Physique* 10 éme Serie, **2**, 22 (1925).
(48) the original derivation of Planck in November 1900 is well described by Pais, Ref. (27).
(49) for example, J. C. Huang, *J. Phys. G, Nucl. Phys.* **14**, 273 (1987).
(50) Pais (Ref. 27) describes this paper, and shows that the Rayleigh-Jeans Law was independently derived by Einstein. He also describes the derivation of this law in the quantum theory by Bose, leading to Bose-Einstein statistics.
(51) E. P. Wigner, *Ann. Math.* **40**, 149 (1939).
(52) L. H. Ryder, *Elementary Particles and Symmetries* (Gordon and Breach, London, 1986).
(53) L. D. Barron, *Physica B* **190**, 307 (1993); M. W. Evans, ibid., p. 310; L. D. Barron, *Molecular Light Scattering and Optical Activity* (Cambridge University Press, Cambridge, 1982).
(54) L. H. Ryder, *Quantum Field Theory*, 2nd edn. (Cambridge University Press, Cambridge, 1987).
(55) E. M. Corson, *Introduction to Tensors, Spinors and Relativistic Wave Equations* (Blackie, London, 1953); M. W. Evans in I. Prigogine and S. A. Rice eds.,

Advances in Chemical Physics Vol. **81** (Wiley, New York, 1992).

(56) M. W. Evans, Vol. **85**(2) of Ref. 15.
(57) as described for example in Ref. 30.
(58) W. Happer, *Rev. Mod. Phys.* **44**, 169 (1972).
(59) B. S. Mathur, H. Tang and W. Happer, *Phys. Rev.* **171**, 11 (1968).
(60) M. W. Evans, *J. Phys. Chem.* **95**, 2256 (1992); also, Refs. 16 and 17.
(61) M. W. Evans, *Physica B* **176**, 254 (1992); **179**, 157 (1992); *Int. J. Mod. Phys. B* **5**, 1263 (1991); *Chem. Phys.* **157**, 1 (1991).
(62) M. W. Evans, *J. Mol. Spect.* **146**, 143, 351 (1991).
(63) in the first paper of Ref. 18, the theory of inverse Faraday effect spectroscopy is developed in terms of the conjugate product.
(64) This monograph, Ref. 27, is notable in that it makes critical reviews of source material, namely the original papers themselves.
(65) Part of the Doctoral Thesis was published as Ref. 47, essentially the key to wave mechanics.
(66) For example, A. Einstein, *Werk. Deutsch. Phys. Ges.* **18**, 318 (1916); *Mitt. Phys. Ges. Zürich* **16**, 47 (1916); *Phys. Zeit.* **16**, 121 (1918).
(67) L. de Broglie, Library of Congress Listings, The National Union Catalog, Pre-1956 Imprints, Vol. 77 (Mansell Information Pub. Ltd., London, 1970), pp. 434-9.
(68) A. Proca, *Comptes Rendues* **190**, 1377; **191**, 26 (1930).
(69) The original papers by Schrödinger, and the reactions of Einstein and contemporaries, are ably described by Pais, in Ref. 27.
(70) See Atkins, Ref. 38, which describes the interpretation by Born which led to the Copenhagen School's view of quantum mechanics. The latter has become predominant. Vigier, in Ref. 2, describes the alternative Einstein-de Broglie interpretation, and recent experimental support for it, see Ref. 72.
(71) P. Ghose, D. Home and G. S. Agarwal, *Phys. Lett. A* **153**, 403 (1991); **168**, 95 (1992).
(72) Vigier (Ref. 2) describes the Copenhagen interpretation, but refers to several recent papers giving support to the Einstein-de Broglie interpretation of quantum mechanics, including: Y. Mizobuchi and Y. Ohtake, *Phys. Lett. A* **168**, 1 (1992). This is a phenomenon in which wave and particle must be simultaneously invoked for its description. It is

References

investigated using a beam splitter composed of two prisms separated by a gap.

(73) well reviewed by S. Kielich and K. Piatek in Vol. **85**(1) of Ref. 15, and in several other reviews of this Special Topical Issue.

(74) M. D. Reid and D. F. Walls, in Vol. **85**(3) of Ref. 15.

(75) P. R. Holland, *Phys. Rep.* **224**, 95 (1993); J.-P. Vigier, Ref. 2.

(76) J. Anandan, *Found. Phys.* **21**, 1265 (1991).

(77) P. R. Holland, *Phys. Rev. A* **46**, R3585 (1992); *Found. Phys.* **22**, 1287 (1992).

(78) Page 126 of Ref. 75a.

(79) described in numerous elementary texts in mathematical analysis.

(80) The original papers, Ref. 68, were later supplemented by A. Proca, *Comptes Rendues* **202**, 1366, 1490; **203**, 709 (1936); *J. Phys. et. Rad.* **7(VII)**, 347; **8(VII)**, 23 (1936).

(81) J. C. Huang, Ref. 49, incorporates finite photon mass into GWS and SU(5).

(82) Y. Aharonov and D. Bohm, *Phys. Rev.* **115**, 485 (1959); see also J. Anandan, *Phys. Lett. A* **164**, 369 (1992).

(83) M. W. Evans in Ref. 15, Vol. **85**(2).

(84) M. W. Evans and J.-P. Vigier, *Found. Phys.* in press.

(85) In Ref. 30, Goldhaber and Nieto ably review experimental evidence and theoretical arguments for finite photon mass.

(86) R. Tanaś and S. Kielich, *J. Mod. Opt.* **37**, 1935 (1990).

(87) J. M. Rauch and F. Rohrlich, *The Theory of Photons and Electrons* (Addison-Wesley, London, 1959).

(88) W. Heitler, *Quantum Theory of Radiation* (Oxford University Press, New York, 1954).

(89) B. L. Silver, *Irreducible Tensor Methods* (Academic, New York, 1976).

(90) D. F. Walls, *Nature* **306**, 141 (1983).

(91) Special Issue of *J. Mod. Opt.* **34**, Nos. 6/7 (1987).

(92) Special Issue of *J. Opt. Soc. Amer.* **4B**, No. 10 (1987).

(93) R. Loudon and P. L. Knight, *J. Mod. Opt.* **34**, 709 (1987).

(94) M. C. Teich and B. E. A. Saleh, *Quantum Opt.* **1**, 153 (1989).

(95) D. F. Walls and P. Zoller, *Phys. Rev. Lett.* **47** 709 (1981).

(96) S. F. Pereira, M. Xiao, H. J. Kimble and J. L. Hall, *Phys. Rev. A* **38**, 4931 (1988).
(97) reviewed by P. D. Drummond in Vol. **85**(3) of Ref. 15.
(98) R. J. Glauber, *Phys. Rev.* **131**, 2766 (1963); reviewed in Ref. 15.
(99) J. Perina, *Quantum Statistics of Linear and Nonlinear Optical Phenomena* (Reidel, Dordrecht, 1984).
(100) M. Schubert and B. Wilhelmi, *Nonlinear Optics and Quantum Electronics* (Wiley, New York, 1986); see also Ref. 15.
(101) B. W. Shore, *The Theory of Coherent Atomic Excitation* Vols. **1** and **2** (Wiley, New York, 1990).
(102) A. Miranowicz and S. Kielich, in Ref. 15, Vol. **85**(3), and other articles in Vol. **85**(1).
(103) R. Tanaś and S. Kielich, Ref. 86, *Opt. Commun.* **45**, 351 (1983); *Opt. Acta* **31**, 81 (1984); reviewed in Ref. 15. The conjugate product to which the third Stokes operator is proportional in this papers has recently (and incorrectly) been asserted to be unobservable by A. Lakhtakia, *Physica B* **191**, 362 (1993).
(104) S. Kielich and K. Piatek, Vol. **85**(1) of Ref. 15. This article discusses squeezed states of light in nonlinear optics, with emphasis on harmonic generation. The standard theory of the conjugate product which is inherent in this treatment appears to be disputed by D. M. Grimes, *Physica B* **191**, 367 (1993), following a paper by A. Lakhtakia, Ref. 103.
(105) P. W. Atkins, Ref. 38.
(106) R. Tanaś and S. Kielich, *Quantum Opt.* **2**, 23 (1990).
(107) W. Heitler, *Quantum Theory of Radiation* (Oxford University Press, New York, 1954).
(108) A. Piekara and S. Kielich, *Arch. Sci.* **11**, 304 (1958), reviewed by R. Zawodny in Ref. 15, Vol. **85**(1). The inverse and ordinary Faraday effects are inter-related by S. Woźniak, B. Linder, and R. Zawodny, *J. Phys. Paris* **44**, 403 (1983).
(109) P. S. Pershan, *Phys. Rev.* **130**, 919 (1963).
(110) G. Wagnière, *Phys. Rev. A* **40**, 2437 (1989).
(111) P. W. Atkins and M. H. Miller, **75**, 491 (1968).
(112) N. L. Manakov, V. D. Ovsiannikov, and S. Kielich, *Acta Phys. Polon.* **A53**, 581, 595, 737 (1978).
(113) predicted by M. W. Evans, Ref. 60a, and verified qualitatively by Warren *et al.*, Ref. 16.
(114) In Ref. 17 the results by Warren *et al.* are disputed by Harris and Tinoco on the grounds of symmetry. Harris and Tinoco (Ref. 17) supply a second order

perturbation calculation of ONMR shifts, a calculation which provides a result many orders of magnitude smaller than observed experimentally in Ref. 16.
(115) in A. A. Hasanein and M. W. Evans Ref. 13.
(116) W. S. Warren, S. Mayr, D. Goswami, and A. P. West Jr., *Science* **259**, 836 (1993).
(117) Ref. 116 is a reply to R. A. Harris and I. Tinoco, *Science* **259**, 835 (1993).
(118) J. Frey, R. Frey, C. Flytzannis, and R. Triboulet, *Opt. Commun.* **84**, 76 (1991); also *J. Opt. Soc. Am.*, B **9**, 132 (1992).
(119) L. de Broglie, *Comptes Rendues* **199**, 445 (1934).
(120) Refs. 54, 68 and 72.
(121) L. de Broglie, The National Union Catalog, A Cumulative Author List 1963 to 1967 (J. W. Edwards, Ann Arbor, Michigan, 1969), Vol. **21**, pp. 118-119.
(122) inter-related by Goldhaber and Nieto, Ref. 30.
(123) R. J. Duffin, *Phys. Rev.* **54**, 1114 (1938).
(124) J.-P. Vigier, *Found. Phys.* **21**, 125 (1991).
(125) P. Garbacewski and J.-P. Vigier, *Phys. Lett.* A **167**, 445 (1992).
(126) A. Kyprianidis and J.-P. Vigier, *Europhys. Lett.* **3**, 771 (1987); N. Cufaro-Petroni and J.-P. Vigier, *Found. Phys.* **13**, 253 (1983); N. Cufaro-Petroni, C. Dewsney, P. Holland, T. Kyprianidis, and J.-P. Vigier, *Phys. Rev.* D **32**, 1375 (1985); P. Holland and J.-P. Vigier, *Il Nuovo Cim.* **88B**, 20 (1985).
(127) F. Selleri, *Quantum Paradoxes and Physical Reality* A. van der Merwe, ed. (Kluwer, Dordrecht, 1990).
(128) P. Grigolini, *Nonlinear Quantum Mechanics* M. W. Evans and J. Moscicki, eds. (World Scientific, Singapore, 1994).
(129) D. F. Bartlett and T. R. Corle, *Phys. Rev. Lett.* **55**, 59 (1985); D. F. Bartlett and G. Gengel, *Phys. Rev.* **39**, 938 (1989); D. F. Bartlett, *Am. J. Phys.* **58**, 1168 (1990).
(130) Upper bounds on photon mass are given in standard tables, and in the reviews by Goldhaber and Nieto (Ref. 30) and Vigier (Ref. 72).
(131) the original paper on this well known effect (Ref. 54) is Y. Aharonov and D. Bohm, *Phys. Rev.* **115**, 485 (1959).
(132) R. G. Chambers, *Phys. Rev. Lett.* **5**, 3 (1960).
(133) T. T. Wu and C. N. Yang, *Phys. Rev.* **12**, 3845 (1975).
(134) L. Bass and E. Schrödinger, *Proc. Roy. Soc.* **232A**, 1 (1955).

(135)   M. Moles and J.-P. Vigier, *Comptes Rendues* **276**, 697 (1973).

INDEX

A_μ
  four-vector A_μ 131, 133
  from the Proca equation 130
  longitudinal and time-like parts
    of 157
  manifest covariant of 135
  physical meaning 120, 133
  vector potential due to 201
$A_\mu A_\mu = 0$ condition 149
Absorption 19
Action 38, 171
  principle of least 171, 179
Admixture equation 156
Aharonov-Bohm effect 67, 120, 122,
    127, 131, 133, 135, 136, 183,
    201
  optical 131, 136, 201
  optical equivalent of 122
Angular momentum 2, 3, 12-14, 16,
    19, 20, 29, 31, 32, 35, 44, 45,
    47, 49-55, 58, 60, 62, 63, 65,
    71, 73, 75, 84, 85, 93-95, 97,
    99-102, 103, 112, 141-144, 162,
    164, 168, 171, 178, 181,
    189-191, 196, 197
  coherent states 99
  commutator 49
  electronic 178
  in four dimensions 62
  in special relativity 189
  intrinsic (spin) 84
  of light 2
  operators 44, 99, 190
  wave function 51
Anti-bunching 89
Anti-field 23, 56
  anti-matter experiments 35
  electromagnetic—$(-A_\mu)$ 57
Anti-matter 23
Anti-photon 26, 35

$\hat{B}^{(3)}$
  geometrical nature of 141
  longitudinal 75, 79, 81
  operator 112
$B^{(1)}$, $B^{(2)}$, and $B^{(3)}$
  classical magnetic fields 73
  Maxwellian fields 72
$B^{(3)}$
  exponential decay in 125
  longitudinal 131
  magnetic field 199
  mechanism at first order 110
  non-zero 120, 139
  physical 61, 131, 180

  potential model 158
  real, in Euclidean space 80
  vacuum value of 109
Baryon number 27
Bilinear products 77
Bohr magneton 107
Boost generator(s) 50, 71, 75-81,
    76, 77, 95, 96 104, 190, 196,
  electric fields 76
  four dimensional 96
  infinitesimal 77
  $\hat{K}^{(3)}$ 79, 81
Bose-Einstein statistics 43, 58
Boson 2, 18, 28, 31, 29, 65, 84,
    100, 143
  massive intermediate vector 119
  operator $\hat{B}^{(3)}$ 100
  (the photon) with mass 65

$\hat{C}$
  operator 22, 27
  parity 28
  symmetry 22
  violation 118
Canonical formalism 134
Cartesian basis 85
Casimir invariant 168
$CdCr_2Se_4$ 111
Charge 27
Charge conjugation 21
Circular
  polarization 20, 112
  unit vectors 17
Circular basis 4, 17, 20, 49, 51,
    61, 71-73, 75, 77-80, 84, 92-
    94, 99, 144, 159, 185
  geometrical commutators 73
  invariance and duality 185
Classical
  Electrodynamics 6, 17
  field $B^{(3)}$ 55, 58
Clebsch-Gordan 86
Commutators
  cyclical basis 73
  Lie algebra 72, 73
  matrices 72, 73
Condition $A_\mu A_\mu = 0$ 67
Conjugate fields 187
Conjugate product 26, 35, 41, 45,
    61, 64, 82, 98, 103-106, 108,
    114, 115, 104, 114, 181, 182,
    201
  free space 181
Conservation

213

laws in electrodynamics 4
of charge 57
Constant of motion 144
Copenhagen interpretation 38, 48
Coulomb gauge 120, 123
Coupling, coefficients 86
$\hat{CP}$ violation 25
$\hat{CPT}$ theorem 24
Cyclical
  Lie algebra 72
  relations 17
Cyclic algebra 136
Cyclically symmetric
  relations 3, 153
  structures between plane wave components 9
Cyclotron frequency 181

D representations 85
D'Alembert equation 5, 56, 58
De Broglie
  Guiding theorem 1
  matter waves 18, 58
  photon equation 117
  wave particle dualism 15
Density of states 19
Dipole moment, magnetic 178
Dirac
  condition 67
  particle and antiparticle 117
  sea 59
Dirac equation 27, 55
  for a free electron 57
  for the free electromagnetic field 57
Distance proportional red shift 123
Double slit experiments 38, 48
Double star motion 123
Dual transformation 10, 33, 193
  field in vacuo, $-i\mathbf{B}^{(3)}/c$ 156
  pseudo four-vector of $F_{\rho\sigma}$ 165
Duffin, Kemmer and Petiau equations 117

Effective four-current 129
Effective free space current four-vector $J_\mu^{(eff)}$ 129
Eigenstates coherent 101
Eigenvalues operator 12
Einstein-de Broglie
  interpretation of dualism 122
  interpretation of light 47
  theory of light 138
Einstein equation of motion 128
Einweg-Welcherweg problem 122
Electric field 2, 10, 23, 34, 35, 60, 76-80, 92, 128, 136, 147--149, 154, 158, 159, 167, 168, 187, 190, 199
  boost generators 76
  discrete symmetries 35
  in space-time 78
  Lie algebra of in the Lorentz group 80
  pseudo four-vector representations 161
Electric field strength
  and amplitude $E^{(0)}$ 23
  of the plane wave 2
Electrodynamic four-tensor 4
Electromagnetic
  density 11, 13
  electromagnetic wave and classical symmetry 32
  field 152, 193
  four-tensor $F_{\mu\nu}$ 141
  phase, ϕ 38
  sector 118
Electromagnetism, four-tensor of 63
Electron
  beams 134
  charge on 22
  lifetime 119
  plasma 181
  radiating 30
Energy 12
Eu^{++} calcium fluoride glass 108
Euclidean E(2) 18
Euler-Lagrange field equations 154
Expectation values 37, 48
  $\mathbf{B}^{(3)}$, of $\hat{B}^{(3)}$ 51
Experimental evidence
  for $\hat{B}^{(3)}$ 103
  for $\mathbf{B}^{(3)}$ 104
  $m_0 \neq 0$, $\mathbf{B}^{(3)} \neq 0$ 120
  prototype ONMR 144
  self interference 122
  survey of data 115

$F^{\mu\nu}$, time-component of 63
Faraday
  nonlinear rotations 113
  optical effect (OFE) 111
  rotation spectrum 111
Fermat principle 39
Fermi-Dirac statistics 58
Fermion 29
Field
  operator $\hat{B}^{(3)}$ 100, 144
  quantization 149
  transverse and longitudinal 71
Field component(s) 60, 71, 143, 155, 156, 158, 165, 196, 201
  $\mathbf{B}^{(1)}$, $\mathbf{B}^{(2)}$, $\mathbf{B}^{(3)}$ 71
Field equations, inter-relation of 55
Finite photon
  mass—experimental indications 68
  radius 67
Fourier
  analysis 89
  integral 89
Four-momentum 13

Index

Free space
  $B^{(3)}$ in 68
  covariance of the Proca equation 126
  E_μ and B_μ 167
  electromagnetism 3
  longitudinal solutions 18
  Lorentz invariance of the Maxwell equations 125
  phase free $B^{(3)}$ 65
  stationary states of one photon 42

Galaxy clusters 123
Gauge 6, 18, 22, 23, 33, 42, 55-57, 59, 60, 62, 62-68, 91, 104, 119-121, 127, 131, 134, 136, 138, 147-152, 154, 170
  fixing 149, 150
  invariance 56, 119, 148
  limiting condition described by $A_\mu A_\mu = 0$ 119
  transformation 57, 119
  quantization of A_μ 60
Generalized position 147
Geometrical basis 17
Glauber coherent state 98
Grand unified theory 119
Group theory 84
Gupta-Bleuler
  condition 60, 149
  field quantization 120
  method 60
Gyromagnetic ratio 178

Hamiltonian
  one photon operator $\hat{H}$ 45
  operator for a single photon 40
Hamilton-Jacobi equation
  of motion 171
  relativistic, e in A_μ 171
Hamilton's principle 39
Harmonics
  scalar spherical 83
  vector spherical 82, 84
Heisenberg uncertainty principle 47, 48, 51, 52, 54, 89, 98, 99, 101, 121, 145
  applied to angular momentum 54
Helicity 31
  free photon 167, 190
  massless photon 166
Hertz potential method 159
Hubble constant 123
Hyperpolarizability 107, 179

$i\hat{B}^{(3)}$, longitudinal 79, 81
Imaginary term 4, 9-11, 33-35, 60, 61, 63, 65, 66, 79-82, 87, 97, 108, 127, 131, 133, 136, 141, 144, 147, 148, 156, 158, 161, 167, 181, 185, 187, 191, 195

Intensity 11
Interference pattern 134
Inverse Faraday effect 3, 11, 12, 14, 19, 34, 35, 45, 61, 65, 71, 97, 103, 105, 106, 110, 140, 141, 181, 183
  magnetization by light 103, 105
  resonance structure 113
Isomorphism 80

Kemmer equation 27, 55
Klein-Gordon equation 42

Lagrange equation 147
Lagrangian field theory 147
  of electromagnetism 147
Laser induced fringe patterns 122
Lepton number 27
Levi-Civita symbol 72
Liénard-Wiechert equations 129
Light 5
  intensity 61, 82, 103, 106, 110, 115, 144, 181-183, 156
  magnetic 106, 108, 181
  mass 68
  quantum hypothesis 17, 37
Light induced shift 109
Light-like condition 67, 147
Light intensity
  antisymmetric part 103, 106
  square root of 115
Light scattering, antisymmetric 82
Light squeezing 52, 54, 89, 98
Linear momentum 8, 12, 44
Little group 170
London equation of superconductivity 128
Longitudinal field(s) 76, 81, 82, 88, 139
  irreducible representations of 82
Longitudinal part of A_μ 66
Lorentz 1, 4, 5, 10-12, 14, 15, 47, 50, 60-62, 64-66, 74-78, 80-82, 91, 95, 103, 104, 120, 121, 124, 126, 131, 141, 143, 147-150, 152, 154-157, 161-166, 168-170, 174, 175, 185-187, 196, 197, 199, 201-203
  condition 61
  gauge 33, 60, 62, 120
  invariants 4, 125
  transformation matrix 162
Lorentz force
  due to $F^{(3)}_{\mu\nu}$ 199
  due to $T^{(3)}_{\mu\nu}$ 199
  equation 164
Lorentz group
  Lie algebra of 50, 74
  magnetic fields in 75

Madelung fluid elements 138
Magnetic circular dichroism (MCD)

111
Magnetic field(s)   17, 22, 33, 34,
    35, 38, 41, 49, 50, 51, 53, 58,
    61-63, 65, 66, 72-76, 79, 80,
    82, 91, 94, 95, 97, 99, 106,
    108, 109, 111, 113, 126, 129,
    130, 136, 142, 160, 161
  $B^{(3)}$ 34
  discrete symmetries 35
  Lie algebra of in the Lorentz group 80
  longitudinal $B^{(3)}$ from Proca equation 129
  of free space electromagnetism 139
  of the electromagnetic plane wave in vacuo 75
  operators 51, 49, 76, 143
  pseudo four-vector representations 161
Magnetic flux density 17
  amplitude $B^{(0)}$ 23
  of an electromagnetic plane wave 2
Magnetizability 107
Magnetization   3, 5, 12, 19, 35,
    68, 71, 97, 103, 105, 107, 108,
    123, 137, 139-141, 156, 171,
    178-183
  by light 3, 97, 171, 180
  due to $B^{(3)}$ 19
  permanent 171, 178
Magneto-optic effects 115
Mass 5
Mass invariant 168
Mass of the universe 123
Matter bridges 123
Matter waves 37
Maxwell
  displacement current 122
  stresses in free space 202
Maxwell equations   27, 33, 34, 42,
    55, 58, 61, 64, 67, 71, 88,
    103, 118, 124, 125, 127-129,
    141, 144-146, 156, 161, 169,
    196, 201
  $\hat{C}$ conservation 55
  inhomogeneous 127
Meissner effect 128, 129
Microwave radiation in the universe 121
Minkowski
  equation 163
  metric tensor 151
  notation 10
  space-time 34
Molecular property tensors 113
Momentum
  conjugate 147
  position, equal time commutator 153
  stress-energy-linear, tensor $T^{(3)}_{\mu\nu}$

201
Motion reversal 21

Neutrino 27
Noether's theorem 6, 56
Non-linear product $B^{(1)} \times B^{(2)}$ 145
Non-locality
  action at a distance 122

Olbers paradox 123
Operators
  $\hat{a}$ 94
  $\hat{a}^{(1)}$ and $\hat{a}^{(2)}$ 100
  $\hat{e}$, $\hat{a}$, $\hat{\jmath}$ and $\hat{B}$ 101
  $\hat{e}^{(1)}$ or $\hat{a}^{(1)}$ 100
  compound irreducible tensor 85
  creation and/or annihilation 54, 62. 63, 64, 77, 89
  field 91
  for infinitesimal rotations 85
  longitudinal $\hat{a}^{(3)}$ 62, 63, 77, 84, 95, 97
  negative frequency 155
  time-like $\hat{a}_0^{(1)}$ 96
  quantum mechanical angular momentum 73
  raising (creation) 84
  unit $\hat{e}^{(1)}$, $\hat{e}^{(2)}$, and $\hat{e}^{(3)}$ 93
Optical fiber 137
Optically induced NMR shifts 3
Optical NMR 108, 139, 181, 183

$\hat{P}$
  conservation 25
  violation 118
Parity
  intrinsic 29
  inversion 21
Pauli-Lubansky   pseudo-vector 162, 165, 189
Pauli matrices 27
Photomagneton   3, 44, 45, 49, 52--
    54, 52, 55, 58, 59, 73, 75, 97,
    98, 103, 112, 141, 145, 171
  $\hat{B}^{(3)}$ 45, 54, 55, 59, 73, 97, 103, 141
Photon   1-6, 8, 9, 11-16, 18-20,
    26-33, 35, 36, 37, 39-45, 47-
    55, 56, 58-60, 62, 63, 65-68,
    71, 73, 82, 84, 89, 91-102,
    103, 104, 112, 117-127, 129-
    131, 136-146, 148-150, 152,
    165-170, 190, 191, 196, 197,
    203, 204
  as massless particle 20
  $\hat{C}$, $\hat{P}$, and $\hat{T}$ 30, 32
  energy of one photon 4
  linear momentum 8
  longitudinal and time-like admixture 153
  low velocity 123

Index

parity of 28
radius 119
rest 5, 204
scalar, vector and spinor fields 42
spin 18
symmetry 29, 30
transverse and longitudinal 71
Photon angular momentum 14, 19, 16, 100
Photon mass 5, 6, 9, 14-16, 18, 67, 68, 97, 104, 117, 119-123, 126, 127, 129, 130, 136-139, 146, 148, 149, 167, 169, 170
and effective current 126
finite 8, 18, 120, 137, 148
invariance with 148
physical A_μ 137
Photon number
average 98
operator 91
Photon operators
annihilation 89, 91
creation 89, 91
longitudinal 93
time-like, $\hat{a}^{(0)}$ 93
Planck law 11, 17, 19
Plane waves 14, 87
Poincaré group 31
Polar tensor antisymmetric 72
Position/momentum 20
Positron, radiating 28, 30
Poynting vector 4
time averaged 11
Proca and d'Alembert equations 65
Proca equation 5, 7, 15, 18, 27, 37, 59, 65-68, 117, 120, 122--124, 126-131, 135, 138, 139, 149, 152, 170
general solutions of 129
Pseudo four-vector 162, 195
Psi function, ψ 38
Pump laser intensity 113

Quantization volume 92
Quantized peaks in the N log z plot 123
Quantum optics 89
Quantum potential
non-locality 121
Quasars 123

Red shift
cosmological 123
Rest mass 203
Rest-volume 203
Rotation generator(s) 44, 45, 49, 50, 72-76, 79-81, 93, 115, 141, 144, 157, 190, 196, 197
four dimensional 96
infinitesimal 73
magnetic fields 76
$\hat{J}^{(3)}$ 81, 141

Rotation group 85
irreducible representations of 83
Rotation operators
infinitesimal, $\hat{J}$ 94

Scalar potential 159
Space-time
four dimensional 72
Lie algebra of electric and magnetic fields in 76
Special relativity
fundamental constant 121
Spin field 10-12, 14-20, 24, 30, 31, 34, 35, 37-39, 41, 44, 51, 53, 60, 61, 71, 72, 81, 93, 103, 113, 120, 124, 136, 171
$B^{(3)}$ 71, 93, 103, 180
$\hat{B}^{(3)}$ 81
Spin invariant 168
Stokes
operators 101, 143
parameters 92
Superluminal action at a distance 121
Susceptibility 179

Tolman's "tired light" 123

Unified field theory 119
gauge invariance 127
Unit vector
$e^{(1)}$, $e^{(2)}$, and $e^{(3)}$ 185
operators 99

Vacuum state 155
Vector
axial unit 72
classical field 85
fields 85
polar unit 77
Verdet constant 108

Wave and particle
physical co-existence of 122
simultaneous existence 122
Wigner 3-j symbols
$B^{(3)}$ 86

Yukawa potential 123

Fundamental Theories of Physics

Series Editor: Alwyn van der Merwe, *University of Denver, USA*

1. M. Sachs: *General Relativity and Matter.* A Spinor Field Theory from Fermis to Light-Years. With a Foreword by C. Kilmister. 1982   ISBN 90-277-1381-2
2. G.H. Duffey: *A Development of Quantum Mechanics.* Based on Symmetry Considerations. 1985   ISBN 90-277-1587-4
3. S. Diner, D. Fargue, G. Lochak and F. Selleri (eds.): *The Wave-Particle Dualism.* A Tribute to Louis de Broglie on his 90th Birthday. 1984   ISBN 90-277-1664-1
4. E. Prugovečki: *Stochastic Quantum Mechanics and Quantum Spacetime.* A Consistent Unification of Relativity and Quantum Theory based on Stochastic Spaces. 1984; 2nd printing 1986   ISBN 90-277-1617-X
5. D. Hestenes and G. Sobczyk: *Clifford Algebra to Geometric Calculus.* A Unified Language for Mathematics and Physics. 1984
     ISBN 90-277-1673-0; Pb (1987) 90-277-2561-6
6. P. Exner: *Open Quantum Systems and Feynman Integrals.* 1985   ISBN 90-277-1678-1
7. L. Mayants: *The Enigma of Probability and Physics.* 1984   ISBN 90-277-1674-9
8. E. Tocaci: *Relativistic Mechanics, Time and Inertia.* Translated from Romanian. Edited and with a Foreword by C.W. Kilmister. 1985   ISBN 90-277-1769-9
9. B. Bertotti, F. de Felice and A. Pascolini (eds.): *General Relativity and Gravitation.* Proceedings of the 10th International Conference (Padova, Italy, 1983). 1984
     ISBN 90-277-1819-9
10. G. Tarozzi and A. van der Merwe (eds.): *Open Questions in Quantum Physics.* 1985
     ISBN 90-277-1853-9
11. J.V. Narlikar and T. Padmanabhan: *Gravity, Gauge Theories and Quantum Cosmology.* 1986   ISBN 90-277-1948-9
12. G.S. Asanov: *Finsler Geometry, Relativity and Gauge Theories.* 1985
     ISBN 90-277-1960-8
13. K. Namsrai: *Nonlocal Quantum Field Theory and Stochastic Quantum Mechanics.* 1986   ISBN 90-277-2001-0
14. C. Ray Smith and W.T. Grandy, Jr. (eds.): *Maximum-Entropy and Bayesian Methods in Inverse Problems.* Proceedings of the 1st and 2nd International Workshop (Laramie, Wyoming, USA). 1985   ISBN 90-277-2074-6
15. D. Hestenes: *New Foundations for Classical Mechanics.* 1986
     ISBN 90-277-2090-8; Pb (1987) 90-277-2526-8
16. S.J. Prokhovnik: *Light in Einstein's Universe.* The Role of Energy in Cosmology and Relativity. 1985   ISBN 90-277-2093-2
17. Y.S. Kim and M.E. Noz: *Theory and Applications of the Poincaré Group.* 1986
     ISBN 90-277-2141-6
18. M. Sachs: *Quantum Mechanics from General Relativity.* An Approximation for a Theory of Inertia. 1986   ISBN 90-277-2247-1
19. W.T. Grandy, Jr.: *Foundations of Statistical Mechanics.*
 Vol. I: *Equilibrium Theory.* 1987   ISBN 90-277-2489-X
20. H.-H von Borzeszkowski and H.-J. Treder: *The Meaning of Quantum Gravity.* 1988
     ISBN 90-277-2518-7
21. C. Ray Smith and G.J. Erickson (eds.): *Maximum-Entropy and Bayesian Spectral Analysis and Estimation Problems.* Proceedings of the 3rd International Workshop (Laramie, Wyoming, USA, 1983). 1987   ISBN 90-277-2579-9

Fundamental Theories of Physics

22. A.O. Barut and A. van der Merwe (eds.): *Selected Scientific Papers of Alfred Landé.* [*1888-1975*]. 1988 ISBN 90-277-2594-2
23. W.T. Grandy, Jr.: *Foundations of Statistical Mechanics.*
    Vol. II: *Nonequilibrium Phenomena.* 1988 ISBN 90-277-2649-3
24. E.I. Bitsakis and C.A. Nicolaides (eds.): *The Concept of Probability.* Proceedings of the Delphi Conference (Delphi, Greece, 1987). 1989 ISBN 90-277-2679-5
25. A. van der Merwe, F. Selleri and G. Tarozzi (eds.): *Microphysical Reality and Quantum Formalism, Vol. 1.* Proceedings of the International Conference (Urbino, Italy, 1985). 1988 ISBN 90-277-2683-3
26. A. van der Merwe, F. Selleri and G. Tarozzi (eds.): *Microphysical Reality and Quantum Formalism, Vol. 2.* Proceedings of the International Conference (Urbino, Italy, 1985). 1988 ISBN 90-277-2684-1
27. I.D. Novikov and V.P. Frolov: *Physics of Black Holes.* 1989 ISBN 90-277-2685-X
28. G. Tarozzi and A. van der Merwe (eds.): *The Nature of Quantum Paradoxes.* Italian Studies in the Foundations and Philosophy of Modern Physics. 1988 ISBN 90-277-2703-1
29. B.R. Iyer, N. Mukunda and C.V. Vishveshwara (eds.): *Gravitation, Gauge Theories and the Early Universe.* 1989 ISBN 90-277-2710-4
30. H. Mark and L. Wood (eds.): *Energy in Physics, War and Peace.* A Festschrift celebrating Edward Teller's 80th Birthday. 1988 ISBN 90-277-2775-9
31. G.J. Erickson and C.R. Smith (eds.): *Maximum-Entropy and Bayesian Methods in Science and Engineering.*
    Vol. I: *Foundations.* 1988 ISBN 90-277-2793-7
32. G.J. Erickson and C.R. Smith (eds.): *Maximum-Entropy and Bayesian Methods in Science and Engineering.*
    Vol. II: *Applications.* 1988 ISBN 90-277-2794-5
33. M.E. Noz and Y.S. Kim (eds.): *Special Relativity and Quantum Theory.* A Collection of Papers on the Poincaré Group. 1988 ISBN 90-277-2799-6
34. I.Yu. Kobzarev and Yu.I. Manin: *Elementary Particles. Mathematics, Physics and Philosophy.* 1989 ISBN 0-7923-0098-X
35. F. Selleri: *Quantum Paradoxes and Physical Reality.* 1990 ISBN 0-7923-0253-2
36. J. Skilling (ed.): *Maximum-Entropy and Bayesian Methods.* Proceedings of the 8th International Workshop (Cambridge, UK, 1988). 1989 ISBN 0-7923-0224-9
37. M. Kafatos (ed.): *Bell's Theorem, Quantum Theory and Conceptions of the Universe.* 1989 ISBN 0-7923-0496-9
38. Yu.A. Izyumov and V.N. Syromyatnikov: *Phase Transitions and Crystal Symmetry.* 1990 ISBN 0-7923-0542-6
39. P.F. Fougère (ed.): *Maximum-Entropy and Bayesian Methods.* Proceedings of the 9th International Workshop (Dartmouth, Massachusetts, USA, 1989). 1990 ISBN 0-7923-0928-6
40. L. de Broglie: *Heisenberg's Uncertainties and the Probabilistic Interpretation of Wave Mechanics.* With Critical Notes of the Author. 1990 ISBN 0-7923-0929-4
41. W.T. Grandy, Jr.: *Relativistic Quantum Mechanics of Leptons and Fields.* 1991 ISBN 0-7923-1049-7
42. Yu.L. Klimontovich: *Turbulent Motion and the Structure of Chaos.* A New Approach to the Statistical Theory of Open Systems. 1991 ISBN 0-7923-1114-0

Fundamental Theories of Physics

43. W.T. Grandy, Jr. and L.H. Schick (eds.): *Maximum-Entropy and Bayesian Methods.* Proceedings of the 10th International Workshop (Laramie, Wyoming, USA, 1990). 1991 ISBN 0-7923-1140-X
44. P.Pták and S. Pulmannová: *Orthomodular Structures as Quantum Logics.* Intrinsic Properties, State Space and Probabilistic Topics. 1991 ISBN 0-7923-1207-4
45. D. Hestenes and A. Weingartshofer (eds.): *The Electron.* New Theory and Experiment. 1991 ISBN 0-7923-1356-9
46. P.P.J.M. Schram: *Kinetic Theory of Gases and Plasmas.* 1991 ISBN 0-7923-1392-5
47. A. Micali, R. Boudet and J. Helmstetter (eds.): *Clifford Algebras and their Applications in Mathematical Physics.* 1992 ISBN 0-7923-1623-1
48. E. Prugovečki: *Quantum Geometry.* A Framework for Quantum General Relativity. 1992 ISBN 0-7923-1640-1
49. M.H. Mac Gregor: *The Enigmatic Electron.* 1992 ISBN 0-7923-1982-6
50. C.R. Smith, G.J. Erickson and P.O. Neudorfer (eds.): *Maximum Entropy and Bayesian Methods.* Proceedings of the 11th International Workshop (Seattle, 1991). 1993
ISBN 0-7923-2031-X
51. D.J. Hoekzema: *The Quantum Labyrinth.* 1993 ISBN 0-7923-2066-2
52. Z. Oziewicz, B. Jancewicz and A. Borowiec (eds.): *Spinors, Twistors, Clifford Algebras and Quantum Deformations.* Proceedings of the Second Max Born Symposium (Wrocław, Poland, 1992). 1993 ISBN 0-7923-2251-7
53. A. Mohammad-Djafari and G. Demoment (eds.): *Maximum Entropy and Bayesian Methods.* Proceedings of the 12th International Workshop (Paris, France, 1992). 1993
ISBN 0-7923-2280-0
54. M. Riesz: *Clifford Numbers and Spinors* with Riesz' Private Lectures to E. Folke Bolinder and a Historical Review by Pertti Lounesto. E.F. Bolinder and P. Lounesto (eds.). 1993 ISBN 0-7923-2299-1
55. F. Brackx, R. Delanghe and H. Serras (eds.): *Clifford Algebras and their Applications in Mathematical Physics.* Proceedings of the Third Conference (Deinze, 1993) 1993
ISBN 0-7923-2347-5
56. J.R. Fanchi: *Parametrized Relativistic Quantum Theory.* 1993 ISBN 0-7923-2376-9
57. A. Peres: *Quantum Theory: Concepts and Methods.* 1993 ISBN 0-7923-2549-4
58. P.L. Antonelli, R.S. Ingarden and M. Matsumoto: *The Theory of Sprays and Finsler Spaces with Applications in Physics and Biology.* 1993 ISBN 0-7923-2577-X
59. R. Miron and M. Anastasiei: *The Geometry of Lagrange Spaces: Theory and Applications.* 1994 ISBN 0-7923-2591-5
60. G. Adomian: *Solving Frontier Problems of Physics: The Decomposition Method.* 1994
ISBN 0-7923-2644-X
61 B.S. Kerner and V.V. Osipov: *Autosolitons.* A New Approach to Problems of Self-Organization and Turbulence. 1994 ISBN 0-7923-2816-7
62. A. Heidreder (ed.): *Maximum Entropy and Bayesian Methods.* 1994
ISBN 0-7923-2851-5
63. J. Peřina, Z. Hradil and B. Jurčo: *Quantum Optics and Fundamentals of Physics.* 1994
ISBN 0-7923-3000-5
64. M. Evans and J.-P. Vigier: *The Enigmatic Photon.* Volume 1: The Field $B^{(3)}$. 1994
ISBN 0-7923-3049-8

KLUWER ACADEMIC PUBLISHERS – DORDRECHT / BOSTON / LONDON